**Books are to be returned on or before
the last date below.**

0 3   ᴘ 1995

Data Acquisition for
Signal Analysis

Companion with this book:

Digital Methods for Signal Analysis, George Allen & Unwin, 1979

Other works by Dr Beauchamp

Signal Processing, George Allen & Unwin, 1973
Walsh Functions and their Applications, Academic Press, 1975
Exploitation of Seismograph Networks (ed.), Noordhoff Press, 1975
Interlinking of Computer Networks (ed.), D. Reidel, 1979

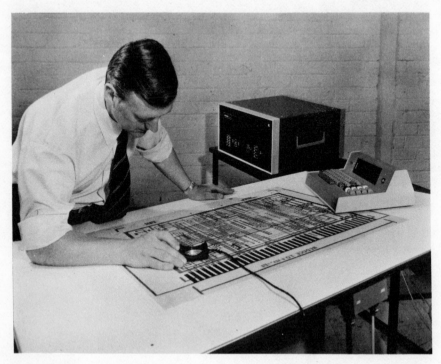

Frontispiece. A curve-follower device (reproduced by kind permission of Ferranti Ltd).

Data Acquisition for Signal Analysis

by

K. G. BEAUCHAMP, C.G.I.A., C.Eng., Ph.D., M.I.E.E.
Director, Computer Services,
University of Lancaster

and

C. K. YUEN, B.Sc., M.Sc., Ph.D.
Lecturer, Information Science Department,
University of Tasmania

London
GEORGE ALLEN & UNWIN
Boston   Sydney

GEORGE ALLEN & UNWIN LTD
40 Museum Street, London WC1A 1LU

© K. G. Beauchamp and C. K. Yuen, 1980

British Library Cataloguing in Publication Data

Beauchamp, Kenneth George
    Data acquisition for signal analysis.
    1. Signal processing — Digital techniques
    I. Title   II. Yuen, C. K.
    621.38'043'02854044   K 80-40663

    ISBN 0-04-621028-8

Typeset in 10 on 12 point Press Roman at the Alden Press Ltd
Oxford, London and Northampton,
and printed in Great Britain
by Biddles Ltd, Guildford, Surrey

CONTENTS

Preface

The past decade has seen rapid advances in the methods and equipment for data acquisition applications, largely due to the further development of digital and integrated-circuit technology, which has made available new techniques and more sophisticated and extremely reliable components as well as systems. With this advance comes the increased need for comprehensive and comprehensible information concerning the way in which such pieces of equipment are to be used, as well as their limitations under particular conditions. The present book is an attempt to bridge this information gap.

Unfortunately, it is often the case that the person carrying out analysis of the data will not have the combination of technical knowledge and practical electronic experience that would be needed to construct systems for the acquisition of scientific data in a manner which is satisfactory for both the measurement and the analysis objectives. As often as not, the task of actually acquiring the data will be delegated, whether intentionally or unintentionally. But even in such cases, both parties will need to know enough about the subject to enable meaningful communication to take place. With the recent emphasis on the microprocessor in relation to 'intelligent' techniques of data acquisition, a development which considerably widens the available choices in systems design and data analysis, the need for a broad range of knowledge becomes all the more critical.

This book aims to supply the range of background and general knowledge required. It is intended to complement the authors' earlier book *Digital Methods for Signal Analysis*, which discusses the data analysis aspects, only leaving the primary problem of data acquisition to the present volume.

We have arranged the eight chapters to provide, first, a treatment of the main components found in data acquisition equipment, namely those for measurement, recording and digital conversion. The components are then discussed as part of complete acquisition systems and the role of the microprocessor as a component in such systems is studied. Also considered are the various auxiliary tasks required in data acquisition, including an introduction to those communication problems that arise when data are accessed over distance and transferred by various means to the recording and analysis equipment.

Inevitably in a wide-ranging topic such as this, encompassing many different disciplines, much relevant information will be left out and a simplified treatment given where the complexity of a particular area might justify otherwise. A notable example is the all-too-brief introduction to microprocessors and their programming and interfacing needs, topics which alone justify many excellent and complete volumes of textbooks. However, we hope that by the inclusion of numerous references at the end of each chapter, this book will succeed in providing an overview of the many problems that beset the successful acquisition of data, and also point the way for further study by the discerning reader.

The authors would like to acknowledge, with grateful thanks, the assistance given to them by very many people and organisations, including Ferranti CETEC Graphics Ltd, Analog Devices Ltd, Bestobell Meterflow Ltd, Bryant Ltd, Endevco Inc., Racall Ltd, and Professor A. Sale of the University of Tasmania, Dr A. G. Piersol of the University of Southern California, Dr G. H. Byford of the RAF Institute of Aviation Medicine, and Dr D. A. Linkens of the University of Sheffield.

K. G. Beauchamp
University of Lancester
C. K. Yuen
University of Tasmania

Chapter 1

Introductory

1.1 Introduction

Computers are tools for the *processing* of information. However, before processing can start, it is necessary to *collect* the data upon which the computer will act. This requires, in the first instance, acquisition of the required information from its source. It is then necessary to convert the information into a particular form suitable for the sort of analysis and manipulation we wish to carry out. This two-step process is called **data collection**. To take an example, the sources of information in a population census are individual persons. Data are acquired from each person and recorded on a suitable form. Unless special arrangements are made to ensure that these forms can be read directly by the computer, it is usually necessary to convert the data into a machine-readable state, by, for example, punching them onto computer cards; but even then the data collection process is not necessarily complete. One may find it too cumbersome to perform all the analysis on all the primary data because of the large amount available. Various tables summarising the information in different ways may be produced as intermediary results for further, specialised analysis. Sometimes the primary data may be discarded once we have produced these intermediary results!

The subject of this book, **data acquisition**, is a restricted kind of data collection. It is of interest mainly to scientists and engineers. There are two distinguishing features. First, the information we wish to acquire is contained in **physical variables**, such as temperature, velocity and electric current. The sources of the information may be humans, animals, plants, inanimate objects or industrial equipment, and to acquire the required information it is necessary to connect some measuring devices to these objects so that the values of the physical variables may be determined. This means that special equipment may need to be designed to convert the output of the measuring devices into a form suitable for computer processing. Second, the physical variables are usually functions of time; indeed, some may change very rapidly with time. In consequence, the data collection process needs to be highly automated in order that the data acquisition equipment can match the information source in speed.

The above discussion explains why data acquisition is a distinct subject from data collection in general. It also explains why the subject tends to be oriented towards engineering, particularly electronics. Much of the measurement and data conversion equipment contain electronic components, and the design of data acquisition systems must involve a great deal of engineering knowledge.

However, because of the widespread use of data acquisition, a certain amount of standardisation has been imposed, and the design process, regardless of the particular context, has been made highly modular. Numerous manufacturers market equipment that plays a well-defined function as part of the data acquisition process, and a well-informed user should have little difficulty in building up a complete data acquisition system with such modules, even though he may know little about their internal construction. It is the purpose of this book to discuss step-by-step the function and characteristics of various components of data acquisition systems to help the reader become such an informed user.

Processing of the data thus acquired forms a subject of its own and is dealt with adequately in many books, including the authors' companion volume to this text [1]. A list of some of these is included in the references at the end of this chapter [2–8].

1.2   Signal Sources and Energy Transfer

In data acquisition applications, any physical variable carrying information is called a **signal**. However, it should be clear that whether a specific variable contains information depends on one's point of view. When one is trying to receive a particular radio station, the transmission from other stations is merely unwanted interference, or **noise**, even though other listeners may well consider these transmissions as signals. Thus there is a virtually unlimited range of physical variables which may, potentially, be signals, with a similarly unlimited range of signal **sources**. We can classify signals in various ways. One scheme is according to their sources; another by the physical nature of the signals themselves. Under the former classification scheme we have:

1. Biological signals: signals produced by plants and animals, such as body temperature and rate of CO_2 production.
2. Physiological signals: signals produced by humans, mainly for medical diagnostic purposes, such as brain-waves (electro-encephalographs) and heart recordings (electro-cardiographs).
3. Environmental signals: signals produced by some collective phenomenon of nature, such as the speed of wind or river flow, sunlight intensity, animal population or density of vegetation in a particular area.
4. Instrumental: signals produced by laboratory or industrial equipment (e.g. vibration, pressure changes, etc.).

The list is not necessarily exhaustive, but does encompass most signals of interest which are processed using computing techniques.

Turning to the physical nature of signals, by far the most important physical variable capable of carrying information is the **electric current** or **voltage** as electric signals are easy to detect, produce and control, and their behaviour is relatively well understood. In consequence, a great many of our measurement devices and pieces of processing equipment are based on electricity, even though other ways of information handling are possible. Some signals are produced by their sources in the form of electric currents, e.g. the electro-cardiograph. Most other signals are related to some form of energy, whether mechanical (force, velocity, acceleration, sound), thermal (temperature), or electromagnetic (light, radiation). As is well known, energy may be converted from one form to another, which is why each of the above variables may be used to produce an electric signal. Microphones, for example, produce electric currents from sound, photoelectric detectors do so from light, and strain gauges from force. Thus, regardless of the form in which signals are initially generated, it is always possible to convert the information into an electric form. That is, given a non-electric signal, we can devise some instrument which will accept this and output an electric current or voltage that reproduces the information in the signal. Such an instrument is called a **transducer**, which is usually the first point of contact between a data acquisition system and a signal source. Very few direct digital transducers exist. Most transducers are designed to convert the physical phenomena under investigation into a continuous (analog) signal and are followed by an analog-to-digital converter to produce an input acceptable to a digital computer.

1.3   Continuous and Sampled Data Acquisition

As we said earlier, many signals vary with time and are generally continuous functions of time, having a distinct value for every different moment, t. Even within a finite duration, there is an infinite number of different moments of time, at each of which a signal may have a different value. Thus, theoretically the complete reproduction of the information in a signal is impossible because we need to acquire an unlimited amount of data. Furthermore, because each value is a continuously varying quantity (i.e. a real number), to record its numerical value with absolute precision we require an unlimited number of digits, which is again impossible.

We see that any process of data acquisition must involve some **approximation**. We must measure a signal at a discrete set of moments, and at each moment the actual value of the signal is approximated by a finite-digit representation. We call the first process **sampling**, the second, **quantisation**. The combined effect is illustrated in Figure 1.1. It is clear that the data acquired will differ from the original signal and thus contain **sampling and quantisation errors**. Despite the disturbing appearance, the effects of sampling and quantisation make no radical difference. Any instrument must have a finite response time, so that it is never possible to reproduce exactly all the instantaneous variations in a signal. Further, signals are inherently always subjected to various sources of interference and contain errors. Sampling and quantisation merely increase the amount of error already present, and the task of designing data acquisition equipment is simply to ensure that, regardless of cause, the total error is within certain tolerances. As

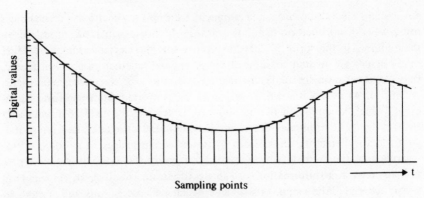

Fig. 1.1   Digitisation of a continuous variable

we shall see, there are well-defined mathematical relations that can be used to estimate the size of sampling and quantisation errors from the properties of the signal and the data acquisition parameters. By selecting the right sampling rate and numerical precision, we can ensure maximum accuracy at minimum cost.

Sampling and quantisation are necessary for data input into computers because they are digital machines, and the whole conversion process is called **digitisation**. Under the current technology, data processing using digital methods is predominant for a number of reasons. First, the reduction in the cost of digital circuits during the past two decades has made it possible to acquire complex equipment at reasonable prices. Secondly, digital methods are flexible. A processing task may be achieved by suitable computer programming, rather than by hardware construction. There are few constraints on the variety of analysis one can carry out and changes are relatively trivial. Finally, digital processing permits higher accuracy at a cost, since we can always attain this by increasing the number of digits and using formulae of greater precision and stability.

At the same time, we must point out that it is possible to perform certain types of processing directly on signals, without first requiring digitisation. Such methods are called **analog data processing**. For example, the car gear-box is really a variable ratio multiplier, which multiplies the input torque (delivered from the engine) by a number selected by the driver to produce the output torque. However, most analog computing is based on electric signals. Besides the well-known processes of arithmetic, it is also possible to differentiate and integrate a function, to compare two functions, and to filter out unwanted frequencies, using analog circuits.

Thus, we may frequently find that analog methods can play a useful part in the data acquisition process, mainly for the purpose of **data reduction** and **signal improvement**. For example, it may be possible to combine a large number of rapidly varying signals to produce a small number of slowly varying signals that contain all the essential information of interest to us. This, then, greatly reduces the complexity of subsequent data acquisition and processing. Also, we find occasionally that the original signals have properties different from those required by our instruments, which cannot handle them as they are. In these cases, we may need to use analog circuits to convert them to a form that matches our equipment. To summarise, we need to analyse carefully the special circumstances of a particular data acquisition problem to determine how to apportion the task between analog and digital techniques.

1.4 Manual Data Entry

A large number of measuring devices, such as thermometers and barometers, produce data in a form suitable for human inspection. The output is usually in analog form, e.g. the length of a column of mercury or the position of a needle above a dial, because this can be easily produced from the input signal and read by the human observer. This then is a simple form of data acquisition: numbers are read off a scale and written down, and the figures are used for subsequent processing. Clearly, this can be practical only for slowly varying signals, as the speed of human reading and recording is quite low. The first improvement one can make is to get the signal recording performed by the *instrument*. For example, the output pointer of the measuring device is connected to a pen, and a roll of paper is pulled under it at constant speed. As the signal varies, the position of the pen changes, and this is recorded on the paper, with the value at each moment recorded at a different position on the graph. Subsequently, readings are taken from the graph. This arrangement resolves the problem caused by the mismatch between the rate of change of the signal and the speed of human recording, since the recording of the signal on the paper and the human reading of the data occur separately, and the latter need not keep up with the former in speed.

However, even here, there is still the physical effort of reading numbers off graphs. The second improvement is the construction of digital meters, which convert the signal value into a number. The number may either be shown on a numerical display for human inspection, or the meter may have a **hard copy unit** to print the number out, saving the human effort to write it down. As the input varies, numerical output is produced repeatedly.

There remains, however, the cost of making the data readable by the computer. If the numbers are merely recorded on a piece of paper, then someone will have to input the data into the machine by typing them into a computer terminal, or punching them onto cards. For large amounts of data this is very costly. The next improvement is to connect the measuring instrument to some output device that produces the data on a medium that can be taken to a computer for direct input. Paper tapes were once widely used for this purpose, though in more recent years magnetic tape, cassette and magnetic disc have become feasible in terms of cost.

Thus, there has been a gradual reduction in the part played by humans in the data acquisition process, with their function progressively taken over by complex instruments of lower cost and better performance. Obviously, the next logical step is to connect the measuring instruments directly to the computer. We shall discuss the advantages of this so-called **on-line** data acquisition in the next section.

1.5   On-line Data Acquisition

When measuring devices are directly interfaced to a computer, we eliminate the need for the intermediary data storage and transfer of data from the acquisition system to the computer. This, however necessary, is not all that is required. It is also usually necessary to provide some method of 'back-up' for the data, so that if mishaps occur during processing and data corruption takes place, one can simply read the data in again from the original computer input medium. The elimination of the medium from the process brings problems as well as advantages. The real reason for on-line data acquisition lies elsewhere.

First, on-line acquisition offers greater speed. The delay in putting data onto the storage medium and in reading them back later is eliminated. This is specially important when data are collected from multiple input channels. In off-line acquisition, the data must usually be mixed together and written onto the same output unit, and the computer needs to unscramble them later into separate streams. A second advantage is that one no longer needs dedicated output devices connected exclusively to the data acquisition equipment. Once the computer receives the data from the measuring instrument, it can write the data on any of its output devices, for back-up, subsequent processing, or transfer to another computer. A third advantage is that more flexible modes of data acquisition become possible. Because the measuring devices are connected to the computer, they can be controlled directly by a suitable computer program, which can choose what, and how much, data are to be collected, depending on various operational conditions. A great deal of human intervention can thus be eliminated, and full automation of data acquisition and processing becomes feasible. Closely allied to this is the possibility of **real-time** data processing and control, so that the computer may, in addition to receiving data, also send output data to industrial or laboratory apparatus to control their operations as a function of the input data. One example is the control of a chemical plant, where information relating to the operation may be sent for immediate analysis to a computer. This determines the actions necessary to maintain efficient operation, such as keeping a boiler at some specified temperature or controlling the content of a particular chemical. A second example may be taken from a missile control system, where the computer receives information about target position and movement from a radar system, and processes it to determine the optimum course for the missile, thus sending the appropriate controlling signals to it.

These are on-line processing activities dependent upon some action being taken as a result of data acquisition. It should be recognised that fully on-line data acquisition systems are complex and costly. The computer must be permanently linked to the input system. This may mean a dedicated processor, such as a microprocessor, although with a time-sharing operating system it is possible to carry out data acquisition at the same time as other processing, such as editing and program development. In either case, the processor must have sufficient speed and capacity to cope with demands made on it. Furthermore, a great deal of specialised work is needed to interface measuring devices to computers, requiring both hardware and software. Because of the variability of signal sources and measuring equipment, special interfacing circuits may need to be developed, and this demands engineering expertise. This means that the cost:benefit ratio of on-line acquisition systems needs to be evaluated very carefully. With certain limited types of on-line systems, it may be possible to link the various measurement and processing units together using a standard type of interface connection, known as a **data bus**. This reduces the task of developing the system to a computer programming operation since the hardware connections are already standardised and require no development. We shall have more to say about this mode of connection in a later chapter.

1.5.1 DATA LOGGING

One particular class of on-line data acquisition system is the multiple-channel **data logger**.

In an industrial process, e.g. a chemical plant, a vast amount of information is recorded in various ways during its operation. Various degrees of data reduction are necessary to convert this mass of data into a usable form. For many purposes, the data deriving from different areas of the plant may need recording (logging), at appropriate intervals, e.g. every hour, or every few minutes, or even continuously. This may be carried out with a simple multi-channel recording device, such as an analog strip-chart recorder or digital recording voltmeter [9]. Often, however, the data will need to be studied only during fault conditions, so that some means of temporary storage is needed to permit print-out of data, upon command, related to current or immediate past operating conditions. For this purpose, the data logger will contain some control logic or a mini/microcomputer and digital memory, so that this and other types of controlled data access can be programmed.

Data logging is often of itself the sole function of a simple computer dedicated to this purpose. No measurement or processing takes place and the recording of information is the 'end-product' of the system. The data logger will accept data relating to plant measurements conveyed electrically as analog or digital signals over a number of communication lines. Some limited conversion into meaningful units may take place before the information is printed out or 'logged'. Where data are logged by recording on computer media such as magnetic tape, they may be transferred easily to an off-line computer for reduction and processing as necessary. This can provide a very effective way of getting information about plant operation to a digital computer. Development of such a procedure may lead to a full-scale computer processing and control system for the plant. In this case, the original data-logging function becomes absorbed in the general process-control system and we have the type of control system described in Chapter 6.

1.6 Visual Data Acquisition and Conversion

A very special sort of data is that of visual information which may need translation into a form suitable for entry into a digital computer. This is a special case because of the large amount of digital data produced related to only a single image and because of the problems of conversion.

The source data is often acquired fairly easily using photographic means. However, conversion equipment and methods can be complex, particularly since much pre-processing may need to be carried out before the data are in the required form for analysis. In later chapters, visual data conversion and systems will be described. However, computer processing, including correction of errors and extraction of specific characteristics of an image, will not be dealt with; this involves the development of specialised software, a description of which is out of place in a book describing data acquisition methods [10, 11].

1.7 Introduction to Computers

It should by now have become clear that data acquisition does not simply mean the collection of data for entry to a digital computer. We need to carry out some operations on the source information and, in many cases, will need to use a digital computer or digital processing concepts in the actual task of acquiring the data. This applies particularly to the repetitive control operations which will accompany acquisition of data in an on-line situation.

Thus, if we are to understand the application of the computer or microcomputer as a mechanism for data acquisition and control we must first become aware of how the computer functions and how we may program it to accomplish the control and acquisition functions we wish to carry out. We will also need to understand the operation, if not the design, of the various logic elements used in or associated with the digital computer. In this section, we intend to provide the basic essentials of this understanding so that the following chapters will be more readily assimilated. It will also serve as introduction to Chapter 7, where we will take these principles further in a discussion of the microprocessor as an important tool for data acquisition work. Neither this chapter nor the book itself represents a comprehensive treatise on the digital computer. We examine only one aspect of this subject here, namely the way in which computers and logic elements are used for data acquisition. A number of references are included at the end of this chapter which deal with computation in its wider context [12–16].

A digital computer consists of four basic elements:

  (i)   an arithmetic/logic unit (ALU)
 (ii)   a program control unit (PLU)
(iii)   a memory subsystem
(iv)   an input–output subsystem.

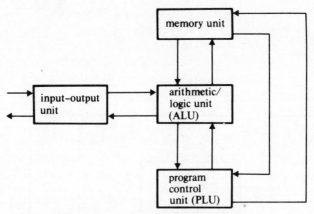

Fig. 1.2   Basic elements of a digital computer

These are interconnected as shown in Figure 1.2.

The ALU forms the essential computing element in which all the basic mathematical and logic function manipulations are carried out, such as adding or subtracting two numbers, comparing two numbers to indicate equality or otherwise, and performing logical operations on pairs of numbers, e.g. OR, AND, NOR, NAND and EXCLUSIVE AND/OR, which we describe later. It will be appreciated that more complex operations such as multiplication, division or decision-making can all be achieved by a suitable series of these basic functions. We need to define how to combine the basic ALU functions in order to form the more complex operations and this definition forms a computer algorithm or **program**.

To control the sequence of operations defined by the program, we need a **program control unit** (PLU). This will include a timing unit or **clock** to maintain a correct sequence of operations and a **program counter** (PC) to keep track of the state of the different sets of control sequences required.

A combination of the ALU and PLU is known as the **central processing unit** (CPU) and we shall see later that a **microprocessor** can be considered as a CPU when combined with other units to form a **microcomputer**.

The **memory** is the largest internal element of a computer and has only two functions, namely to store programs and to store data.

The final element, the **input–output** subsystem, enables the computer to link with the outside world by providing **interfaces** to manual input devices (terminals or papertape units), other input devices (analog-to-digital converters, data loggers) and output devices (printers, visual display units, or equipment linked via digital-to-analog converters).

The essential feature of the digital computer shown in Figure 1.2 is that it is a **stored program** device. Thus, we can work out a series of elementary functions which, when executed one after the other, will perform a given operation such as adding a series of numbers together. This series of functions can be stored in the computer's memory so that by initiating the start of the series it is possible to perform the operation automatically without human intervention.

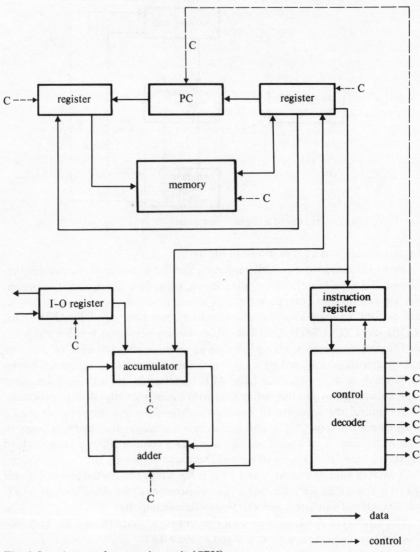

Fig. 1.3 A central processing unit (CPU)

We will now look at the interaction between the CPU and the stored instructions contained in the memory in a little more detail. Figure 1.3 shows a diagram of a typical CPU. In addition to control circuits and arithmetic units, we see that this contains a number of units called **registers** which form an important element in the functioning of the CPU. Registers are temporary storage elements containing a single number or group of binary bits. Some of these are for general use and others have dedicated uses such as the **accumulator** or the program counter which we met earlier. The accumulator usually stores one of the operands to be manipulated by the ALU. A typical operation or instruction would be to direct the ALU to add the contents of some other register to the contents of the accumulator and to store the results in the accumulator itself. Thus, the accumulator behaves both as a source and destination register.

Instructions are stored in the memory and, in order to determine the next action in a program, the CPU will need to examine the contents of the memory in the correct order. Each of the locations within the memory is numbered uniquely and the number that identifies a memory location is known as its **address**. The correct sequence of operations is maintained by always arranging that the address of the *next* instruction to be followed is contained in the **program counter** (PC). The CPU updates the PC by adding 1 to the counter each time it fetches an instruction from the memory. Programmers therefore need to store their instructions in numerically adjacent addresses so that the lower addresses contain the first instructions to be executed and higher addresses contain later instructions. This rule is broken only when the last instruction in a particular group of instructions tells the CPU to jump to a new address and to recommence sequential addressing from this new location. The purpose of this jump instruction will be considered later in some detail. For the present, we may note that this programmed deviation from an orderly sequence of consecutive instructions provides a way of making a decision to carry out a fresh set of new instructions depending on some criteria the programmer can include in his program.

We earlier described the memory as storing instructions or data in a number of locations, each defined by a numbered address. Each of these locations contains a number of binary bits forming collectively one computer word. The computer's **word-length** is determined by the size of its registers and the number of interconnecting paths, which are known as **buses**. The microprocessor we shall be considering in Chapter 7 has an eight-bit word-length and since eight bits are said to constitute a **byte** we refer to this as a byte machine. Other larger machines may have a word-length of several bytes of information.

A typical instruction word may have a format like that shown in Figure 1.4, where the word-length is divided into an **operation code** (opcode or instruction code) and **operand** sections. The opcode tells the processor what to do. The operand or operands contain the addresses of the data to be operated upon by the opcode. The processor fetches an instruction from the memory in two distinct operations. First, it transfers the address in its program counter to the memory; secondly, the memory returns the addressed word to the processor. The CPU stores the instruction word in the **instruction register** (IR) and uses it to control operations during the remainder of the instruction execution.

The mechanism whereby the processor translates an opcode into a specific processing action is contained in the decoding circuits shown in Figure 1.3.

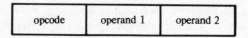

Fig. 1.4   An instruction word format

Various logic circuits are activated and the appropriate output line carries a controlling level to the associated control circuits.

Data is stored in registers and in the memory in the form of a **data word**. This can represent numbers or alphanumeric data. A widely used code for the latter is the American Standard Code for Information Interchange (ASCII). Numbers are represented in binary code. Various types of coding are used, including **binary**, **binary-coded-decimal**, **octal** and **hexidecimal** (base 16) code. We will be discussing some of these coding systems in a later chapter.

Integer numbers can be stored in memories and registers with a sign bit in the first position to show whether the remining binary bit positions contain a positive or negative number. Other representations of signed integers are used and a common system is known as **2's complement** and will be encountered in many microprocessor systems. Here a binary 0 is appended above the highest order magnitude bit as a + sign with the magnitude bits unchanged:

$$\text{binary } 5 \; = \; 0101$$

$$\text{2's complement form} \; = \; 00101$$

For a negative number, a binary 1 is appended above the highest order magnitude bit as a $-$ sign and the magnitude bits replaced as follows: (a) replace all 1's by 0's and vice versa; (b) add 1:

$$\text{binary} - 7 \; = \; -0111$$

$$\text{replacing 1's by 0's, etc.} \; = \; 1000$$

$$\text{add 1} \; = \; 1001$$

$$\text{add minus sign bit} \; = \; 11001$$

This representation may appear complex but provides the advantage that the same hardware equipment can be used both for addition and subtraction.

Arithmetic and logical operations on the stored data are carried out within the CPU in response to control signals decoded from information contained in the instruction register. In the next section, we will be looking at some of the logic elements that carry out these operations. We will also meet these same elements in other chapters, since logical operations are carried out for control purposes whenever digital signals are transferred or manipulated in any way.

1.7.1 LOGIC ELEMENTS

In this section, we shall be developing some of the rules of digital operations which follow a set of basic **rules of logic**. They have their own language expressed in a special form of two-level mathematics known as **Boolean algebra** and are depicted graphically by a set of symbols representing different forms of logical operations.

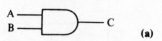

(a)

A	B	C
0	0	0
0	1	0
1	0	0
1	1	1

(b)

Fig. 1.5   The AND operation
   (a) Logic symbol
   (b) Truth table

The fundamental logic operations are AND, OR, and NOT.

The AND operation is shown in Figure 1.5a. The two inputs to this logic element, A and B, must *both* be present and have a logical 1 value in order that the output, C, may also have a logical 1 value. In Boolean algebra, the multiplication symbol of ordinary algebra represents the AND function so we may write

$$C = A \cdot B \tag{1.1}$$

Figure 1.5b shows a table of different combinations of 0's and 1's possible for A and B and the resultant logical state for C. This is known as a **truth table** and completely expresses the operation of this logic element.

The OR operation is shown in Figure 1.6a. Here, *either* of the two inputs A or B may assume a logical 1 value to result in a logical 1 output at C. We express this in Boolean algebra as

$$C = A + B \tag{1.2}$$

The truth table for a two-input OR operation is given in Figure 1.6b.

These AND and OR logic elements can, of course, have more than two inputs where the rules of logic similarly apply. Thus, an n-input AND logic element will require all its n inputs to be at the logical 1 level before a logical 1 level is available at the output.

The NOT operation means simply the logical inversion of an input. Thus a logical 0 is produced for a logical 1 input. We show this by the symbol given in Figure 1.7a. The small circle indicates logic inversion and the triangle represents a non-inverting amplifier having a gain of unity. We can use this logic inverting circle in conjunction with the AND or OR inputs or outputs; several examples

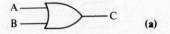

(a)

A	B	C
0	0	0
0	1	1
1	0	1
1	1	1

(b)

Fig. 1.6   The OR operation
(a) Logic symbol
(b) Truth table

A ——▷o—— Ā

NOT   (a)

A
B ——{ NAND }o— C   A
B ——{ NOR }o— C

$C = X.Y.\bar{Z}$

X
Y
Z ——{ }——

A
B ——{ }—— $C = A + \bar{B}$

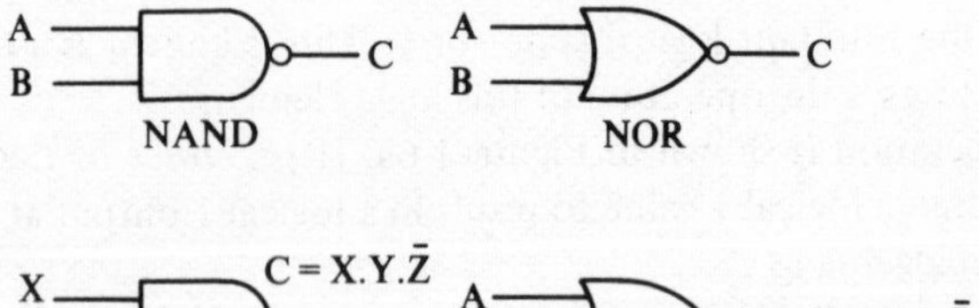

(b)

Fig. 1.7   The NOT operation
(a) Logic symbol
(b) Examples of inverted logic operations

are given in Figure 1.7b. Where the inversion is associated with the output of these logic elements they are given special names of NAND and NOR. In practice, we find that NAND and NOR elements have wider use than AND and OR elements, since they are more versatile and readily implemented by transistor circuits.

We represent the NOT operation in Boolean algebra by adding a bar above the logic character being inverted. Thus for the NOT element of Figure 1.7a we would write

$$C = \bar{A} \qquad (1.3)$$

Two input NAND and NOR operations would be written respectively as

$$C = \overline{A \cdot B} \qquad \text{and} \qquad C = \overline{A + B} \qquad (1.4)$$

As an example of the interconnection of logic elements we consider the operation of binary addition. We wish to add two numbers, A and B, which are each four bits in length, say 1011 added to 0101. The first step is to add the least significant bits, in this example, two 1's. This addition results in a sum term, S, and a carry term, C, to be added to the next column. A truth table for this part of the addition (the least significant bit) is shown in Figure 1.8. Let us consider this for the S column alone. The results of this column occur frequently in digital logic, so that a separate term has been assigned to it called the EXCLUSIVE-OR. Expressed in Boolean terms we write

$$S = \bar{A} \cdot B + A \cdot \bar{B} \qquad (1.5)$$

From the truth table, Figure 1.9a, we note that the output may only be 1 when A *or* B are at the digital 1 level and that the output is zero when A and B are alike. This can be achieved by the interconnection of the three logic elements we have used so far, as shown in Figure 1.9b. The same results can be achieved in other ways, and the logic symbol shown in Figure 1.9c is used to represent this operation without defining the method of connection for the internal logic. (Note the symbol $\oplus$, used to describe the EXCLUSIVE-OR operation.)

The complete operation of adding two binary bits to produce a sum, S, and a carry, C, in accordance with the truth table of Figure 1.8 is obtained by the interconnection of an EXCLUSIVE-OR element and an AND element, as shown in Figure 1.10. This is known as a HALF-ADDER operation.

All the mathematical operations required in a digital computer can be obtained by interconnection of the logic elements described above. Counting and

A	B	S	C
0	0	0	0
0	1	1	0
1	0	1	0
1	1	0	1

Fig. 1.8   Truth table for least significant bit addition

A	B	S
0	0	0
0	1	1
1	0	1
1	1	0

(a)

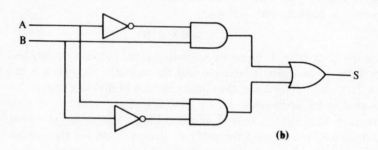

(b)

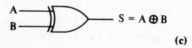

(c)

Fig. 1.9   EXCLUSIVE-OR operation
   (a) Truth table
   (b) Realisation using logic elements
   (c) Logic symbol

decision operations require a digitally controlled switching element or bistable circuit, of which the most widely used is the **flip-flop** (FF). This is a circuit having only two stable states and means for changing the state in response to a digital signal. The basic FF element is shown diagrammatically in Figure 1.11a with its truth table in Figure 1.11b. This is known as a **set–reset FF**. The FF can be set into either output condition Q or $\bar{Q}$ by connecting either S to 1 and then returning S to 0 or by connecting R to 1 and then returning it to 0. It is usual to define the *set* condition as Q = 1, $\bar{Q}$ = 0 and the *reset* condition as Q = 0, $\bar{Q}$ = 1. These changes in output level are seen clearly in the **timing diagram** of Figure

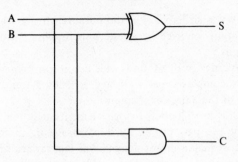

Fig. 1.10   The half-adder operation

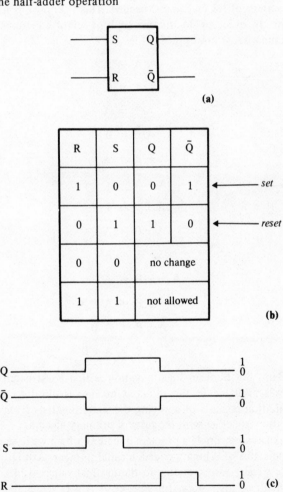

Fig. 1.11   The RS flip-flop
     (a) Logic symbol
     (b) Truth table
     (c) Timing diagram

1.11c, which shows what happens at the outputs when a set pulse is applied to the appropriate input followed by a reset pulse applied a short time later to the reset terminal input.

Whilst this basic design of FF finds wide use in control and computing equipment, other more complex types are found in many systems and we shall need to describe some of these in later chapters.

We conclude this introduction to logic elements and their interconnection by describing briefly the part played by Boolean algebra in solving problems of logical operation. A major aim of designers of logic systems is to achieve a solution to a logical control or calculation problem by reducing to a minimum the number or variety of logic elements needed. This is known as **minimisation by Boolean logic**. In order to do this we apply a set of equations or logic rules which we summarise without proof below.

$$A \cdot A = A$$
$$A \cdot \bar{A} = 0$$
$$A \cdot 1 = A$$
$$A \cdot 0 = 0$$
$$A + A = A$$
$$A + \bar{A} = 1$$
$$A + 1 = 1$$
$$A + 0 = A$$
$$\bar{\bar{A}} = A$$
$$A + A \cdot B = A$$
$$A + \bar{A} \cdot B = A + B$$
$$A + B = \overline{\bar{A} \cdot \bar{B}}$$
$$A \cdot B = \overline{\bar{A} + \bar{B}}$$

$$(1.6)$$

In any problem, the Boolean logic equation is first constructed by combining the logic levels available in the required manner to produce a single output value. Minimisation consists of applying the simplifications given in (1.6) to the Boolean equation, together with the rules of ordinary algebra.

We will take as an example a problem in coding where we have as input values the four logic states of a binary-coded-decimal number, ABCD, where each bit carries a code weight corresponding to the decimal values, 1248. We require to generate from these logic states for any set of ABCD values the corresponding code value for another binary-coded-decimal code having certain special properties, known as **excess-3 code**. We will consider the logic required for a single bit of the excess-3 code, namely the generation of the most significant bit, N, which requires for its calculation the state of all four bits ABCD of the initial coded number. The Boolean equation to do this is given by:

$$C \cdot \overline{B} \cdot A + C \cdot B \cdot \overline{A} + C \cdot B \cdot A + D \cdot \overline{A} + D \cdot A = N \qquad (1.7)$$

If this were implemented directly, we would require three 3-input AND elements, two 2-input AND elements, two NOT elements and one 5-input OR element. But (1.7) can be rearranged as

$$C \cdot \overline{B} \cdot A + C \cdot B(\overline{A} + A) + D(\overline{A} + A) = N \qquad (1.8)$$

and from (1.6) $\overline{A} + A = 1$, so that

$$C \cdot \overline{B} \cdot A + C \cdot B + D = N$$
$$= C(\overline{B} \cdot A + B) + D = N \qquad (1.9)$$

Also from (1.6), $B + \overline{B} \cdot A = B + A$, so that (1.9) simplifies to

$$C \cdot B + C \cdot A + D = N \qquad (1.10)$$

This shows that the code conversion required for each bit of the code can be simplified to the use of two 2-input AND elements and one 3-input OR element, thus representing a considerable saving in logic hardware.

The symbols we use for logical operations in this book conform to an international standard known as ANSI Y32.14 (1973) which finds general acceptance, although readers may encounter different symbols in the literature [17].

1.8 Remote Sensing and Telemetry

In the concluding part of this chapter, we introduce the problems of remote data acquisition and look at the interaction between the computer and telecommunication which is beginning to produce such significant effects. In many data acquisition problems, the signal source is located at some distance from the observer. Here we can distinguish between two classes of problems. In one, the measuring device is located at the source but the equipment to collect and process the data is at another location. This is the case, for example, when we wish to connect monitoring devices to animals, or to control space vehicles. The problem here is to devise systems that will transmit and receive data acquired by the measuring device to the processing equipment. In the second class of problems, the measuring devices are located away from signal sources such as, for example, a satellite information system. The difficulties found here are in the design of long-distance detecting equipment. The former situation is called **telemetry**, while the latter is **remote sensing**. A third situation is found which is a more elaborate form of telemetry. Here we have on-site measuring devices connected to local computers, but the data gathered are collected in a central computer for processing.

Despite the great variety of remote data acquisition problems, there is one design consideration that is common to all of these, namely the signal-to-noise ratio. Because the information has to travel over distance, it is subject to a great deal of interference, and the signal has to be extracted from the interfering noise. It is not sufficient simply to install a sensitive receiver to measure the signals, since a more sensitive receiver will pick up more noise as well.

Nor is a more powerful transmitter necessarily the answer, since there are limitations to the size and weight of remote equipment. Further, in remote sensing we are not able to control the signal strength at all. The answer to these problems lies in appropriate processing, both at the transmitting and at the receiving ends. This is because signals and noise have very different average properties, and by detailed processing, involving analysis of a long piece of received signal, the information can be extracted. For example, the noise received by a remote sensor tends to fluctuate rapidly. If we observe a constant signal source over an extended period of time and accumulate the received signal energy, the noise tends to average out. To further improve signal-to-noise ratio, astronomers have designed radio aerials that accumulate reception from a number of separate detectors, causing further noise cancellation. When the transmission end is also under our control, we will deliberately design the transmitted signal to possess certain regular patterns. Since noise interference disrupts such patterns, it can be detected and rectified. This is the method of error-correcting codes, discussed in Chapter 8.

1.9   Principles of Data Acquisition Systems

We are now ready to take an overall look at data acquisition systems. The first point of contact of such a system with a signal source is the transducer, which measures some physical value and converts its value into an analog electric voltage. This must then be digitised by the processes of sampling and quantisation. However, either before or after digitisation, there may be the need to perform some **pre-processing** in order to achieve data reduction and signal improvement. Analog methods are applied to signals before digitisation, digital methods following digitisation. Sometimes we wish to store the signals temporarily on some medium for later processing, as described earlier. If this is done after digitisation, the medium will be one of the common computer input–output devices. It is also possible to store the analog signals undigitised on tape recorders or similar devices. After some or all of the above intermediary steps, the data are finally entered into a computer.

Thus, we can identify the following parts of a data acquisition system:

   (i)   transduction
  (ii)   pre-processing
 (iii)   storage
 (iv)   digitisation
  (v)   computing,

and these are considered in individual chapters. In Chapter 6, we show examples of data acquisition systems in various contexts.

As mentioned earlier, data acquisition systems are complex and themselves make use of computers to initiate and control the acquisition process. This procedure has become more common with the advent of the microprocessor, which allows the control operations to be carried out efficiently and cheaply. Data acquisition systems, particularly input—output systems, that are able to organise the collection or display of data in this way are often termed 'intelligent devices'. Several examples are described in Chapters 4 and 6.

We conclude this introduction with some general remarks on system design principles. As with other integrated systems consisting of modular components, in data acquisition we must be conscious continually of the need for compatibility, whether in terms of speed, accuracy, or perhaps some other physical and operational characteristics. It is obviously impracticable to measure fast signals using slow transducers, and wasteful to do the reverse. Further, it is important to carry out preliminary analysis to discover where the essential information is contained within the available signals, in order to design systems which will acquire only the minimum necessary amount of useful information. This can be done by selecting transducers of the appropriate speed and sensitivity, by removing unwanted parts of the signals at the pre-processing stage, by choosing the sampling rate to be just high enough to avoid loss of essential information and by selecting a quantisation accuracy just high enough to prevent the introduction of serious errors. Finally, the processing capability, memory size and input—output rates of the computer should be suitably chosen. We should also ensure that the output of each stage of the system has the sort of characteristics demanded by the next stage, e.g. voltages that are within the operational range.

However obvious the need for compatibility may seem to be, failure to achieve full compatibility, either between system components, or between the system design and the actual problem context, remains a main cause of unsuccessful design of data acquisition systems. The value of compatibility considerations cannot be over-emphasised.

To repeat what we have said before, although data acquisition devices are engineering systems, one need not be a professional engineer to be able to engage in their design. What one does need is some understanding of the function and characteristics of data acquisition components, as well as a great deal of common sense. This is the essential message of this book.

References

1. BEAUCHAMP, K. G. and YUEN, C. K. *Digital Methods for Signal Analysis*. George Allen & Unwin, London, 1979.
2. BENDAT, J. A. and PIERSOL, A. G. *Measurement and Analysis of Random Data*. Wiley, New York, 1966.
3. LYNN, P. A. *The Analysis and Processing of Signals*. Wiley, New York, 1973.

4. GRIFFITHS, J. W. R., STOCKLIN, P. L. and SCHOONEVALD, C. VAN. *Signal Processing*. Academic Press, London, 1973.
5. OPPENHEIM, A. V. and SCAFER, R. W. *Digital Signal Processing*. Prentice-Hall, Englewood Cliffs (NJ, USA), 1975.
6. GOODYEAR, C. C. *Signals and Information*. Butterworth, London, 1971.
7. RADER, C. M. and GOLD, B. *Digital Processing of Signals*. McGraw-Hill, New York, 1969.
8. RABINER, L. R. and RADER, C. M. *Digital Signal Processing*. IEEE Press, New York, 1972.
9. *Specification and Selection of Data Logging Equipment*. Engineering Equipment Users Association Handbook 28, Constable, London, 1968.
10. ANDREWS, H. C., TESCHER, A. G. and KRUGER, R. P. Image processing by digital computer. *IEEE Spectrum* 9 20–32, July 1972.
11. ROSENFELD, A. *Picture Processing by Computer*. Academic Press, New York, 1969.
12. EADIE, D. *Introduction to the Basic Computer*. Prentice-Hall, Englewood Cliffs (NJ, USA), 1968.
13. HOLLINGDALE, S. H. and TOOTHILL, G. C. *Electronic Computers*. Penguin, London, 1970.
14. WELLS, M. *Computing Systems and Hardware*. Cambridge University Press, Cambridge, 1976.
15. FORSYTHE, A. I. *Computer Science*. Wiley, New York, 1969.
16. RICE, J. K. and J. R. *Introduction to Computer Science*. Holt, Rinehart & Winston, New York, 1969.
17. IEEE Standard 91–1973 *Graphic Symbols for Logic Diagrams*, ANSI Y32–14. IEEE, London, 1973.

Chapter 2

Data Measurement

2.1 Introduction

In the previous chapter we referred to the correspondence between the physical variable being measured, its energy content and translation into a form capable of being measured and recorded using the digital computer.

Here we shall consider in detail the translation of various forms of **signal energy** into an **electrical signal**. In a later chapter, the conversion of this electrical signal into digital form will be discussed.

Devices used for the measurement of physical quantities in terms of an electrical output signal are called **transducers**. The classification is a broad one, and it is the intention here to consider a fairly limited range of applications which will, it is hoped, cover the main requirements of system measurements as are generally understood in the signal processing environment. Choice of transducers to detect and translate physical energy will depend upon the form of the energy, accuracy requirements, period of operation, environmental conditions, dimensions and cost. These considerations will form the framework of our discussion concerning transducer types which follows.

2.2 Transducers

Considered within the context of this book, transducers may be defined as devices which will translate physical input quantities into electrical output signals. Although we will be considering light transducers later, we exclude from this chapter devices for handling complete images, as these are more properly considered in Chapter 5.

Nearly all transducers are designed to produce a continuous (analog) output. Transducers producing directly a digital output are not easily found, since there do not seem to be any natural phenomena in which some detectable characteristic changes at discrete intervals as a result of a change in temperature, pressure, etc. An exception is the shaft encoder converting angular motion directly into a digitally coded output. Consideration of this, however, will also be deferred to Chapter 5, where analog-to-digital conversion techniques are discussed.

All the transducers described in this chapter produce a continuous signal output. We expect them to have an accurately known input—output characteristic under a given set of conditions so that meaningful properties can be ascribed to the physical process being measured. From the wide range of possible physical measurements, three basic quantities can be recognised:

1. Displacement: linear and angular measurements representing stress, strain, pressure, thickness, force, angle of incidence, etc.
2. Acceleration: linear and angular measurements representing vibration, impact, inertia, etc.
3. Velocity: linear and angular measurements representing speed, rate, momentum, etc.

Other characteristics such as light, heat, and time may also be recorded either as a direct electrical analog or in terms of one of the basic quantities given above.

A variety of physical effects can be utilised in the design of practical transducers, and exhaustive treatment of their design and characteristics is not possible within the confines of this chapter [1–5]. All that we can hope to present is a classification of the main types in use for signal processing purposes, together with sufficient details of their characteristics and use to be of value when assessing the reliability of measurements made using such devices. A broad classification of the physical effects used is stated below.

1. Energy conversion transducers, where energy is abstracted from the system under measurement and converted (with some energy loss) into an equivalent electrical form.
2. Passive transducers, in which a measured change in the physical quantity merely causes a corresponding change in some electrical quantity (e.g. resistance, capacitance, inductance, etc.), and which require an auxilliary source of energy in order to produce an output signal.
3. Feedback transducers: these are characterised by a feedback loop which effects an equilibrium between the input physical quantity and that of an opposing electrical quantity. The force necessary to achieve this equilibrium gives a measure of the physical quantity being measured.

A sub-classification widely used to define transducer types is based upon the particular physical effect utilised in their operation. Figure 2.1 shows the division of these types in terms of energy conversion or passive control action. The third type, that of equilibrium feedback, can utilise one or more of any of these effects in its operation.

2.3   Measurement of Displacement

The most commonly found device for measurement of displacement is the **strain gauge** [4]. The amount of strain is measured by its effect on a long, thin wire attached to a structure. Referring to Figure 2.2, we define strain as

$$\text{strain (m/m)} = e = \frac{\delta L}{L} \tag{2.1}$$

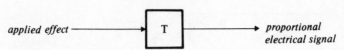

(a) Energy-conversion type:

electro-magnetic, piezo-electric magneto-strictive, thermo-electric, photo-electric

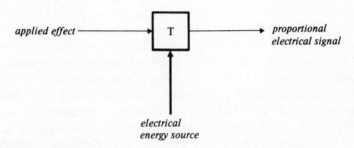

(b) Passive-control type:

resistive, inductive (reluctance), capacitive, mechanical strain (resistive), semi-conductive, thermo-resistive (thermocouple), photo-resistive, hall effect, radioactive

Fig. 2.1   Types of transducers

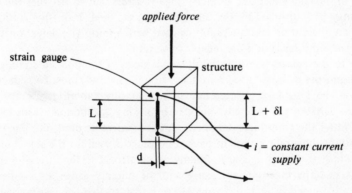

Fig. 2.2   A thin-wire strain gauge

from which the stress, defined as force per unit area, can be obtained from the product of e and Young's modulus of elasticity. The cross-sectional area of this wire is very small (typically $0.0001\,\text{mm}^2$) so that the change in length, δL is reflected in a measurable change in its electrical resistance, δR. For most metals a constant gauge factor can be derived as

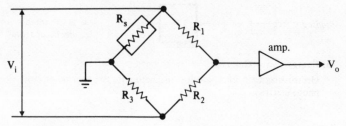

Fig. 2.3   A strain-gauge unit

$$K = \frac{\delta R/R}{\delta L/L} \tag{2.2}$$

Since K is reasonably constant over a wide range of values for strain, a linear relationship between δL and the measured output voltage, $\delta V_0 = \delta R \cdot i$ can be obtained. Modern resistive strain gauges are made of thin foil rather than a circular wire conductor, which eases the problems of manufacture and results in more robust design. The foil is bonded to a thin insulating backing and the bonded strain gauge affixed to the structure. The measurement is obtained by direct transmission of the strain through the backing material to the gauge. This is incorporated in a bridge configuration as shown in Figure 2.3. Changes in resistivity, caused by the alteration in length of the gauge under the applied strain, will unbalance the bridge and cause a potential to be developed across it which is linearly proportional to strain. For a given value of strain, the change in resistivity of foil gauges is fairly low. For this reason other strain-gauge transducers often take the place of the bridge resistors shown, so that the output signal may be multiplied by the number of gauges used. In many practical situations, an alternating current supply is used with appropriate detection arrangements, since this leads to more reliable operation.

Due to the risk of fracture, the dynamic range of the bonded strain gauge is limited and the permissible current density (and hence V_i) must be kept small to minimise the resistance changes due to temperature variations. Compensation for these variations is generally included by means of a thermally sensitive resistor shunting the transducer terminals. An alternative method of compensation is to mount the strain gauges in pairs. One gauge is used for the actual measurement, whilst the other acts as a dummy gauge affixed in the direction of minimum strain. The measurement gauge and the dummy gauge are connected in series and situated in opposite arms of the bridge so that the resistance changes due to temperature effects will tend to cancel.

The magnitude and direction of strain can be determined from a group of similar bonded gauges, often three in number, and referred to as strain-gauge rosettes. The constituent gauges are accurately manufactured on a common base to lie at convenient and accurately known angles to each other. Thus the principal strains and stresses and their angular orientation with respect to the gauge angles can be computed from the three measured strains, allowing a small correction for transverse sensitivity.

The low sensitivity of the wire resistance gauge has led to the development of a form of resistance gauge known as the thin film strain gauge. A ceramic film is vacuum-deposited on to the sensing element and acts as an insulating base. Upon this is deposited, again by vacuum methods, a film of resistive or semi-conductive material through a mask to give a bridge form having the correct strain orientation. A high-resistance gauge is obtained which can withstand high applied voltage and hence produce a higher output for a given power dissipation.

A development of this deposition technique is the fabrication of a pressure diaphragm and strain-gauge bridge with temperature-compensating and signal-conditioning elements on one piece of single-crystal silicon for use in fuel metering and control systems [6].

A special form of vacuum-deposited strain gauge has been developed for building into catheters of very small diameter, which are required for medical purposes. These gauges typically comprise a metal plate $50-200\,\mu m$ thick, supporting within its area a diaphragm which may be as thin as $5\,\mu m$. The gauge itself may be of metal oxide composition giving a gauge factor of about 15. It is essential to use an a.c. bridge with these designs to avoid the effects of electrolytic corrosion, since the catheter will, of necessity, have to be operated within the warm, high-humidity conditions of the body.

Calibration presents a problem with mass-produced thin film and other miniature transducers, since it is not realistic to provide each gauge with its own individual calibration certificate. Instead complete transducer units are manufactured to a given accuracy and considered as interchangeable within ±2% of their specified output values.

A considerable improvement in sensitivity is obtained by the use of semiconductor filaments in place of thin resistance wire or foil. The gauge factor, K for silicon or germanium filaments is of the order $100-200$, compared with a value of about 2 for copper—nickel alloys. To a large extent, the advantage in terms of sensitivity of the semi-conductor gauge is now eroded by the availability of cheap integrated circuit amplifiers which can be used with the resistance strain gauge. The disadvantages of the semi-conductor gauge are its poor linearity, noise and sensitivity to temperature changes.

A resistive displacement transducer which is considerably more robust than the resistance strain gauge is the potentiometric transducer shown in Figure 2.4. This couples the motion caused by pressure directly to the arm of a rectilinear potentiometer through which a constant current is passed, thus affecting the position of the wiper and hence the amplitude of the voltage developed between this and one end of the potentiometer. When connected to a high-input impedance amplifier the arrangement is highly linear and suitable for measuring large pressure differentials. However, it is not suitable for the measurement of very small displacements such as strain, owing to the mechanical limitation of the linkage arrangement.

Other passive methods of displacement measurement utilise a change in inductance or capacitance value. There are several forms of variable inductance transducer, two of which are shown in Figure 2.5. The variable self-inductive

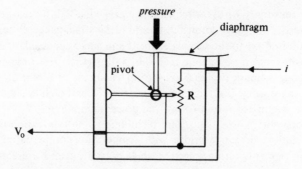

Fig. 2.4   A potentiometric pressure transducer

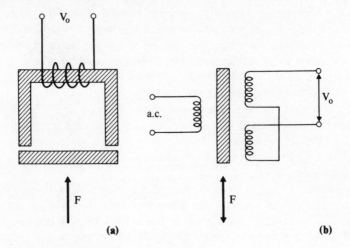

Fig. 2.5   Variable inductance transducers
(a) Self-inductive type
(b) Differential type

type (a) gives an inductance change proportional to pressure over a limited range. The resolution is good and the transducer can be used for the measurement of both static and fluctuating pressures.

The second method (b) uses the change in mutual coupling between two coils which occurs due to movement of their common core. This is known as the differential transformer transducer. Whilst the sensitivity of this arrangement is high, a linear relationship between displacement and the output voltage can only be obtained over a very small part of the total core movement ($\simeq 5\%$). An alternating current energy source is required and an amplitude-modulated output signal is obtained.

The transducers considered above are all passive devices. An energy-conversion transducer which is in wide use is the **piezo-electric strain gauge** (Figure 2.6). The piezo-electric material is subject to mechanical stress and generates a surface

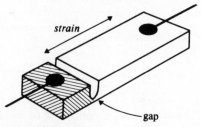

Fig. 2.6   A piezo-electric strain gauge

charge due to deformation of the crystal lattice. This charge can be made to leak away through an external resistance, thus producing a voltage proportional to the imposed stress, since it decreases with time as the charge escapes. The duration of this voltage, however, is finite, so that it can be used only for dynamic measurements. The output voltage produced is given by

$$V = \text{het volts} \tag{2.3}$$

where, h is the piezo-electric strain coefficient (V/m), e is the strain (m/m) and t is the active material thickness (m).

The active material used in early makes of piezo-electric transducers was Rochelle salt or barium titanate. These materials have now been replaced by lead zirconate titanate, due to its high sensitivity. The linearity of the piezo-electric transducer is found to be excellent and the sensitivity high. It finds wide use for the detection of short-term shock strains. Unlike many other displacement transducers, the piezo-electric strain gauge is unaffected by temperature over a very wide range (typically -200 to $+450°$C) and can be used under fairly rigorous conditions.

2.4   Measurement of Acceleration and Vibration

We are often interested in the measurement, not simply of the extent of displacement of an object, but its movement and rate of change of movement, namely acceleration.

If a body is moving in a straight line and its distance from a fixed point on that line is x, then x will change with time t, and its velocity, v, will be represented by dx/dt and its acceleration, a, by d^2x/dt^2. Thus, using the simple relationships $v = dx/dt$ and $a = dv/dt$, we can obtain any desired quantity of motion by differentiating or integrating the signal obtained.

The dimensions of acceleration are length/time^2 and the SI unit is the metre per second square (m/s^2). However, a measure widely used in publications on accelerometers is g, the acceleration due to gravity on the Earth's surface at sea level. This is approximately 9.81 m/s^2, and is used principally as a convenient practical standard for the calibration of accelerometers.

2.4.1 SEISMIC ACCELEROMETER

The general principle of the seismic-mass accelerometer is contained in the equation

$$\text{force} = \text{mass} \times \text{acceleration} \tag{2.4}$$

A given mass, known as the seismic mass, is subject to the acceleration it is desired to measure. This seismic mass is connected to some form of spring and the force on the mass makes it move against the spring until the spring force balances the acceleration force. The displacement of the mass then gives a measure of acceleration. This is shown in Figure 2.7. The frame or transducer housing is fixed to the body whose acceleration is to be measured. The mass is linked to a secondary transducer measuring displacement, such as we met in the previous section, and the displacement of the mass is converted into an electrical signal.

However, there is a problem associated with the simple model of a seismic accelerometer shown in Figure 2.7. A steady acceleration will not produce a proportional steady displacement of the seismic mass. As the housing is accelerated, there is a delay before the seismic mass can follow this acceleration. However, when it reaches its correct position where the force of the spring balances the force of acceleration, it does not immediately stop because of its inherent inertia; consequently, an overshoot occurs. This increases the force of the spring, which attempts to restore the mass to its correct position, but as the mass falls back it would once again overshoot the correct position.

Thus, oscillation occurs due to these interactions between the moving mass

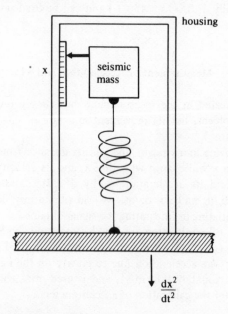

Fig. 2.7   The seismic-mass accelerometer

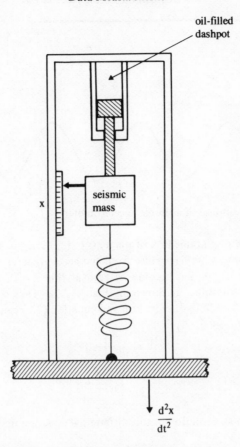

Fig. 2.8   Damping the seismic-mass accelerometer

and the restraining force of the spring. These oscillations can be made to reduce very rapidly by the addition of a controlled amount of extra damping. This is usually carried out by means of a dashpot, consisting of a piston connected to the seismic mass and moving in an oil-filled cylinder. This is shown in Figure 2.8. To see what this damping action needs to do, we consider first the undamped mass-spring accelerometer shown in Figure 2.7. Here, y is the displacement of the housing and hence of the body to which the accelerometer is fastened. The displacement of the seismic mass is x, so that the net displacement is y − x. The net acceleration is given by the second-order equation

$$\frac{d^2(y-x)}{dt^2} = \frac{d^2y}{dt^2} - \frac{d^2x}{dt^2} \tag{2.5}$$

From equation (2.4) and stating the spring restoring force as kx we have

$$m\left(\frac{d^2y}{dt^2} - \frac{d^2x}{dt^2}\right) = kx \tag{2.6}$$

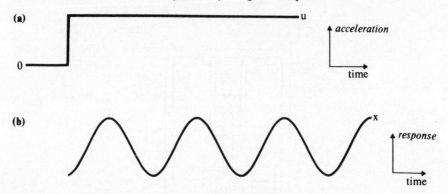

Fig. 2.9   Step response of a simple accelerometer

To find the actual displacement, x of the mass in terms of the acceleration of the housing, we assume a step response from the accelerometer, that is we let the acceleration, d^2y/dt^2 of the housing be zero at time $t = 0$, when it suddenly acquires a constant value, u shown in Figure 2.9a. Thus we have for $t < 0$, $d^2y/dt^2 = 0$ and for $t > 0$, $d^2y/dt^2 = u$ and equation (2.6) can be written

$$m\left(u - \frac{d^2x}{dt^2}\right) = kx \tag{2.7}$$

or

$$\frac{d^2x}{dt^2} = u - \frac{k}{m}x \tag{2.8}$$

It can be shown that a solution to this differential equation may be written as

$$x = \frac{mu}{k}(1 - \cos \omega_n t) \tag{2.9}$$

which represents the displacement for a constant acceleration of u, where $\omega_n = k/m$.

The variation of x with time is shown in Figure 2.9b. This is seen to be sinusoidal by repeating each time $\omega_n t = 2\pi$, so that it is seen to have a frequency $f_n = \omega_n/2\pi$. This frequency, f_n, is called the **natural frequency** of the transducer.

If we now add an extra force equal to $c(dx/dt)$ acting to oppose this oscillatory motion we need to restate equation (2.6) to include this damping term, thus

$$m\left(\frac{d^2y}{dt^2} - \frac{d^2x}{dt^2}\right) = kx + c\left(\frac{dx}{dt}\right) \tag{2.10}$$

or

$$\frac{d^2x}{dt^2} + \frac{c}{m} \cdot \frac{dx}{dt} + \frac{k}{m} \cdot x = u \tag{2.11}$$

Solution of this equation shows that the displacement varies with time in a

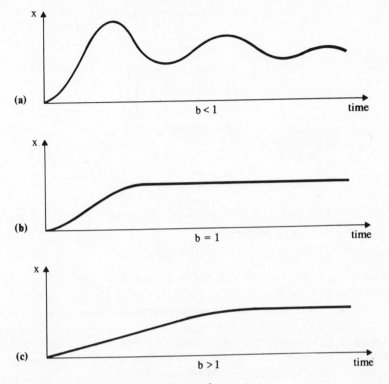

Fig. 2.10   Step response of a seismic accelerometer
(a) Damping less than critical (b < 1)
(b) Critical damping (b = 1)
(c) Damping more than critical (b > 1)

manner dependent on the magnitude of the damping constant, e, the mass, m, and the spring constant, k. This dependence is expressed in terms of a **damping factor**, b

$$b = \frac{c}{2\sqrt{mk}} \tag{2.12}$$

If b < 1 the oscillations are damped (Fig. 2.10a); if b = 1 the oscillations are critically damped (Fig. 2.10b); if b > 1 there is no oscillation (Fig. 2.10c).

In all cases, the displacement settles down to the **steady state** response of the accelerometer, x = mu/k, after a given period of time.

2.4.2   PIEZO-ELECTRIC ACCELEROMETER

An important accelerometer, widely used for impact testing, is the piezo-electric accelerometer, shown in Figure 2.11.

A seismic mass provides pressure on the face of the crystal, proportional to the acceleration magnitude. The voltage developed across the crystal is dependent

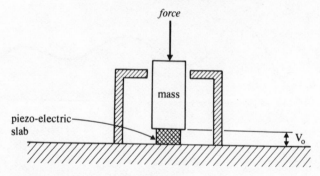

Fig. 2.11   The piezo-electric accelerometer

on the generated charge, Q, and the shunt capacitance of the crystal, plus the cable capacitance, C_p, and is given by

$$V = Q/C_p \text{ volts} \qquad (2.13)$$

The sensitivity is high and the unit can be designed for a very small size, but, as with the pressure transducer, the operation is confined to high frequencies. In fact, they are suitable only for the measurement of dynamic strains and their calibration requires suitable dynamic equipment.

Owing to its favourable sensitivity:mass ratio, this accelerometer is, however, widely used for impact testing. The problem with this form of testing is that very high acceleration values are experienced, and these can rise to as high as 100 000G in a few microseconds. As a consequence, the effective weight of other forms of transducer can reach values of several tons. With miniature piezo-electric accelerometers, the increase is limited to a few hundred pounds and, also due to the small mass, the resonant frequency is generally much higher than most structural elements upon which they are mounted. For similar reasons, such transducers are ideal for the measurement of high-velocity shock transients. Under these conditions, the poor long-term zero stability, and sensitivity to temperature variations are of secondary importance.

Piezo-resistive accelerometers utilise the change in resistivity obtained when the element is subject to a bending stress. As with wire-resistive gauges, they are used in pairs mounted in a bridge configuration.

Recent forms of piezo-accelerometers incorporate a solid-state impedance converter within the body of the transducer, resulting in output impedances as low as 100 Ω. They can therefore deliver a significant amount of power due to this favourable voltage:current ratio.

2.4.3   VIBRATION TRANSDUCERS

Vibration measurements are carried out to detect resonances in components intended for use in locations subject to shock or vibration [7–9]. This may be done by placing the vibration transducer close to the component in its working location or alternatively by affixing the component to a controlled vibrating

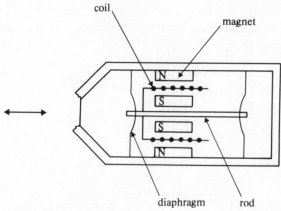

Fig. 2.12   A vibration transducer

mass and driving this mass over a range of frequencies whilst using transducers to measure the resulting vibration. Resonance detection is important for indicating the onset of fracture, or for many other reasons associated with the desired operation of the component being measured.

A common vibration transducer is shown diagrammatically in Figure 2.12. Two spring diaphragms are clamped to a housing and linked by a rod. This carries a hollow cylindrical form around which is wound a coil of copper wire. This assembly is the seismic mass which moves between the poles of a permanent magnet affixed to the housing. When the transducer is vibrated above its natural frequency the relative displacement gives a measure of vibration **amplitude**, whilst the voltage induced in the coil by its movement within the poles of the permanent magnet is proportional to its **velocity** of movement. This voltage is given by

$$e = Blv \text{ volts} \tag{2.14}$$

where B is the magnetic field, l is the total length of the coil, and v is the relative velocity of the coil moving at right-angles to B.

2.4.4   FEEDBACK TRANSDUCERS

Feedback techniques permit the design of transducers which have good linearity and a wide range of application. These techniques were introduced in Section 2.2. In order to achieve an equilibrium between the applied force and an opposing electrical voltage, some form of sensing device is included to control the magnitude of the opposing voltage. This can be a subsidiary feedback loop detecting velocity, rather than linear movement, to permit a rapid equilibrium, and hence higher frequency response.

The advantages of feedback transducers over other methods are increased accuracy, linearity, and sensitivity, and a closer control over natural frequency and damping. A good example of a feedback accelerometer is the Endevco design illustrated in Figure 2.13. This consists of a small, pendulum-pivoted mass

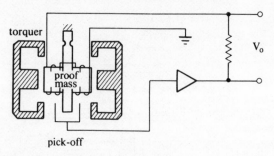

torquer

proof mass

pick-off

Fig. 2.13   The Endevco feedback accelerometer

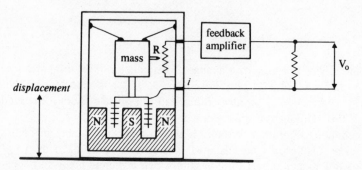

Fig. 2.14   A seismic transducer

constructed of quartz, which is free to move within very small gaps maintained between the mass and the fixed side reference plates of a capacitative pick-off device. The mass also carries with it a coil moving within a permanent magnetic field. Movement of the mass is sensed by means of the change in capacitance at the base of the pendulum. This causes a small change in current which may be applied through a servo-amplifier circuit to cause a restraining force to be given by the coil. Equilibrium is quickly reached and the current maintained through the coil is then a measure of the acceleration obtained. Since a large amount of loop gain is employed, the actual motion required for the mass is extremely small, so that its weight and size can be quite small.

Another example of a feedback transducer is the seismic transducer shown in Figure 2.14. This differs from the normal seismometer in that the coil is no longer a generator, but is provided with a driving current related to the potentiometric displacement value, and hence position of the mass. The current is in such a direction as to oppose any movement of the mass by means of a counter-force acting on the coil. The equilibrium current flowing in the feedback path is thus a measure of the original displacement of the mass.

Feedback systems used to measure pressure may employ a combination of a pressure-sensing diaphragm linked to a differential transducer or a capacitance transducer. The signal produced by the pressure variation is amplified by a servo-system and used to restore the diaphragm to its original position. As with the seismic transducer, the value of the restoration force is a direct measure of the input quantity (in this case, pressure).

2.5   Measurement of Angular Velocities

We measure the rotation of a shaft or linear velocity of a body by means of velocity transducers.

By far the most common velocity transducers are those relying on the generator effect for their realisation. This is shown in Figure 2.15, where the peak value of the induced e.m.f. across the loop rotating within a magnetic field, B, is given by

$$E = AnBv \text{ volts peak} \tag{2.15}$$

where A represents the area of the loop, n is the number of turns and v is the loop angular velocity.

The measurement of angular displacement velocity is made by means of a **tachometer**, which is a compact form of a dynamo generator having the position of the magnet and coil reversed, i.e. the magnet is caused to rotate within a fixed coil. Several windings are included spaced about an axially rotating permanent magnet. The induced e.m.f. in the windings provides a rectangular output waveform having a value proportional to the product of the total number of turns in the stator, n, and the rate of change flux, ϕ, and hence velocity, viz.

$$e = -n\frac{d\phi}{dt} \tag{2.16}$$

Another form of angular velocity measurement utilises a series of small magnets attached to the rotating body. These cause a small e.m.f. to be induced in a

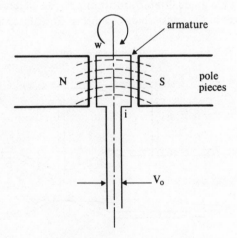

Fig. 2.15   Dynamo velocity transducer

fixed loop adjacent to the rotating shaft with each rotation of the shaft. A train of pulses will be produced, the frequency of which will be proportional to shaft speed. By using a pair of such loops, accurate measurements of axial and radial displacement of the shaft can also be obtained.

2.6   Fluid Flow Measurements

The measurement of flow rate plays an important part in those industries deal- ing with the production and distribution of gas and petroleum products, in the chemical industry, and in water management and the treatment and disposal of sewage. The word 'fluids' here means gases, liquids and mixtures of solids and liquids. Gases differ from the other fluids in that they are compressible, so that the volume of a given mass of gas varies with the applied pressure.

Three types of measurement can be distinguished for fluid flow:

(i)   **velocity of flow** – the *velocity* with which a fluid is moving along a duct
(ii) **volumetric flow rate** – the *volume* of fluid moving along a duct in unit time
(iii) **mass flow rate** – the *mass* of fluid moving along a duct in unit time.

Either differential pressure or mechanical methods are used for fluid flow measurements. The basic principle of the differential pressure method lies in the conversion of some of the kinetic energy of the flowing fluid into potential energy and measuring this as a difference in pressure at two points along the flow path.

A method of measuring volumetric flow rate used since the beginning of the 19th century is that of the Venturi tube. This is shown in Figure 2.16. The shape of the tube is such that a gradual decrease in diameter along the length of the pipe is followed by a gradual return to its original diameter. As a result of these mechanical changes in the constrained channel of flow, there is a drop in pressure

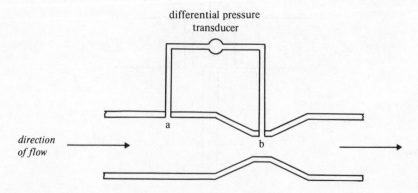

Fig. 2.16   The Venturi tube

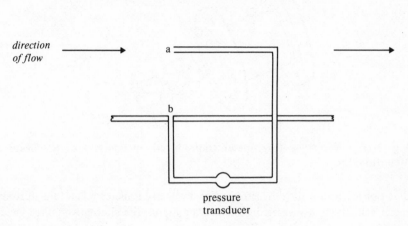

direction
of flow

a

b

pressure
transducer

Fig. 2.17   The Pitot tube

across two pick-off points which can be measured by means of a **differential pressure transducer** and this is simply a specially designed displacement transducer of the type discussed earlier.

A method of measuring velocity of flow is the Pitot tube shown in Figure 2.17. This is particularly applicable to the measurement of air or gas flow. The fluid moving down the pipe again meets two pressure probes, the first lying along the axis of the pipe and the second fitted to the side of the pipe. As with the Venturi tube, a differential pressure difference is set up between these two probes which is measured in the same way. The fluid velocity is given as

$$v = \sqrt{\frac{2(p_i - p_s)}{d}} \qquad (2.17)$$

where $(p_i - p_s)$ is the pressure difference and d is the density of the fluid.

A direct measure of flow velocity is made with the turbine flow meter [10, 11]. This is typically designed to be a free-spinning, multiblade rotor mounted in a non-magnetic cradle. Fluid passing through the meter causes the rotor to turn and a magnetic pick-up coil situated on the body registers a pulse each time the blade interrupts the magnetic flux. A frequency-to-d.c. converter is used for conversion to an indication of fluid flow rate and, of course, this output can be converted to digital form for computer entry. The construction of a turbine flow meter may be seen in Figure 2.18 (which is reproduced by courtesy of Bestobell Meterflow Ltd).

An entirely different method of recording liquid fluid flow is the ultrasonic flowmeter. Two sound detectors are strapped to the outside of the pipe carrying the fluid or, in some cases, immersed within the fluid stream. The detectors are situated on either side of the pipe at an angle to the direction of flow. Sonic pulses are sent alternatively on a line between the detectors, but in opposite

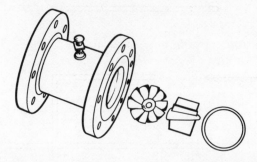

Fig. 2.18   A turbine flow meter (reproduced by courtesy of the Bestobell Meterflow Ltd.

directions. Because the downstream signal velocity is increased and the upstream signal velocity is decreased by the moving liquid, the alternative pulses yield a frequency difference. This difference, when converted to a d.c. current, provides an accurate indication of the average flow velocity. It is also independent of changing fluid parameters such as temperature, pressure, sound, speed, viscosity or density. As with other continuous measurement methods, it is, of course, suitable for digital conversion and on-line monitoring by digital acquisition methods.

2.7   Light Transducers

Two forms of device are in use to translate energy in the visual or infra-red spectrum into electrical energy. The first of these are the **photosensitive resistive** devices. In these, the characteristic of the material used is changed by the action of light falling upon it. An example is the light-dependent resistor or photo-conductive cell. This is usually constructed of cadmium sulphide or lead sulphide, both of which have the property of changing their electrical resistance in accordance with the intensity of light falling upon them. A change in resistivity of several orders of magnitude is obtained for quite moderate illumination variations, so that the method is a very sensitive one.

The second set of devices are those relying on the change of resistivity occurring within a semi-conductor. These give rise to the photodiode and photo-transistor, which may be incorporated directly into signal conditioning and output electronic circuits.

Photodiodes are light sensitive devices which may be either pn junction diodes or npn phototransistors. The former have a more linear characteristic and are faster in reaction to light. For many purposes, such as position sensing, where a high sensitivity is required, the greater output of the phototransistor would be used.

2.8 Acoustic Transducers

Conversion of acoustic energy into electrical energy by means of a microphone is too well known to justify a large place in this chapter.

A simple and common form of transducer is the carbon-granule microphone. This relies on the slight variations of resistivity that occur through a mass of loosely packed carbon granules when the surface of the granule mass is subject to the arrival of acoustic waves. However, this is a fairly noisy device, having a low upper frequency response.

An alternative is the moving-iron microphone. This depends on the movement of an iron or steel diaphragm, upon which the sound waves impinge, to induce a signal in a detection coil mounted in a magnetic field. This too has a poor frequency response, due to the resilience of the diaphragm, although it is less noisy than the carbon-granule device.

Improved performance is obtained from the moving-coil microphone. This operates by moving a coil attached to a cone and diaphragm which move within a uniform magnetic field. Acoustic waves impinging on the cone cause movement of the assembly and induce corresponding current changes in the coil. Although this is made light in weight, the mechanical resilience of the mounting again limits the upper frequency response to about 7 kHz.

Two microphone designs having much improved upper frequency responses are the ribbon and crystal microphones. In the former, a corrugated aluminium foil strip mounted with the field of a permanent magnet receives acoustic signals, and vibrates in a lateral direction. This movement induces a small current in the strip in a similar way to the moving-coil microphone. Owing to the very light weight of the foil strip, a good frequency performance can be obtained.

The crystal microphone uses a thin quartz crystal responsive to sound pressure waves falling upon it. A small piezo-electric e.m.f. is generated as described earlier in this chapter. The sensitivity of this microphone is quite high and its frequency response is almost as good as that of the ribbon microphone.

2.9. Miscellaneous Energy Measurements

Other measurements which may be encountered are those concerned with temperature and physical characteristics such as moisture content, pH (chemistry), vacuum measurement and those found in biological measurement [12]. It is only possible to consider these briefly here, although the subject of temperature measurement is of sufficient importance to warrant a more complete treatment.

2.9.1 TEMPERATURE MEASUREMENT

Heat is conveyed from place to place by means of conduction, convection and radiation. All these effects are found in temperature transducers and the correct choice of transducer is dependent on the location and range of temperature it is

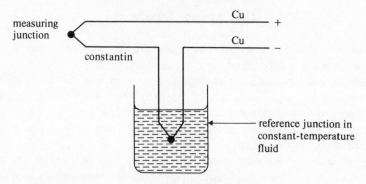

Fig. 2.19   Temperature measurement by thermocouple

required to measure. We will be concerned only with electrical methods here, and non-electrical devices such as bi-metallic and liquid thermometers will not be discussed.

The most common form of thermo-electric transducer is the **thermocouple**. Temperature is measured by conduction at the junction of two dissimilar metals. A second junction is usually connected in series with the first and is kept at a given reference (cold) temperature (Figure 2.19). Measurement utilises the **Seebeck effect**, which defines the appearance of a voltage across a junction when this is subject to a temperature gradient. A reverse effect, the **Peltier effect**, introduces some non-linearity into the process. This defines a change in temperature which results when an electrical current is caused to flow across the junction. For this reason the thermocouple is associated with high-input impedance methods of measuring the e.m.f. across the junction, in order to minimise the current flow.

Many such metal/metal combinations are in use; common examples are platinum/rhodium, iron/constantin, and copper/constantin. Table 2.1 shows the characteristics of a number of these.

Series-connected thermocouples in which a number of hot junctions are mounted on top of one another are known as **thermopiles**. The total electrical output is the sum of the individual e.m.f.'s and a much more sensitive instrument results. Sensitivities of the order of 1 mV/°C are achieved.

Measurement of very high temperatures is achieved by means of a pyrometer which consists of a thermocouple so arranged that it detects the source of heat by radiation only and is removed from actual contact at the source of heat.

Table 2.1   **Thermocouple comparison table**

+Element	−Element	Range,°C	Sensitivity, µV/°C
tungsten	rhenium	0 to +2000	20
chromium	aluminium	−250 to +1400	30
rhodium	platinum	0 to +1500	10
iron	constantin	−250 to +1000	40
copper	constantin	−250 to +400	40

Temperature measurement may also be made by the change in electrical resistance of conductors and semi-conductors when subject to a temperature gradient. By incorporating the heat-measuring conductor in an electrical bridge circuit, extremely good accuracy, sensitivity and stability of the measuring method can be achieved over a range of temperatures from 10 to 1300 K. A coil of wire is wound non-inductively onto an insulating former and the coil is enclosed in a refractory protective sheath. Copper leads are connected from the coil to a Wheatstone bridge.

The change in resistance for conductors with temperature is given by

$$R = R_o(1 + \alpha t + \beta t^2) \tag{2.18}$$

where R_o is the initial resistance and α and β are constants.

Semi-conductors are also used for this purpose, and, in this case, they are known as **thermistors**. These have a negative temperature coefficient and exhibit a temperature–resistance relationship of the following exponential form

$$R/R_o = \exp \beta \left(\frac{1}{T} - \frac{1}{10} \right) \tag{2.19}$$

where R is the resistance at absolute temperature, T, R_o is the resistance at absolute temperature, T_o and β is a constant.

The most common semi-conductor used for this purpose is germanium, due to its stability and lack of drift from calibration measurements.

Calibration of these temperature-measuring devices can be made against the onset of freezing conditions for various substances solidifying from the molten state. At this point, their temperature remains constant until complete solidity is obtained. This is known as primary calibration. Secondary calibration involves direct calibration with a previously calibrated instrument and is the more usual procedure.

2.9.2 HUMIDITY AND pH MEASUREMENTS

A modification of the technique of wet- and dry-bulb thermometry is used for the measurement of moisture content by a fairly simple method. The moisture-laden airstream is caused to flow over the surface of a thermocouple. Relative humidity may then be deduced from differences between the indicated temperature and ambient temperature. A chemical method involves the use of lithium chloride, which is a highly hygroscopic material. The absorbed water content is evaporated by electrodes energised from an alternating current supply, thereby heating the resultant solution. An equilibrium condition is realised for a constant voltage supply between the rate of water evaporation and absorption. A measure of the solution resistance then gives an indication of the moisture content.

An electronic method of measuring moisture content is to expose the capacity plates of a tuned circuit to a material absorbing the moisture, which becomes the dielectric of the tuned circuit. The change in oscillator frequency is a function of the moisture content.

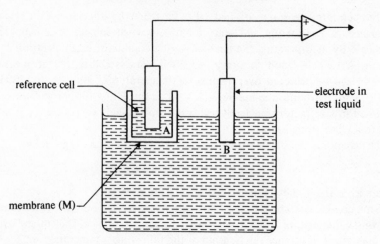

Fig. 2.20   pH measurement

Measurement of pH is being increasingly used in all types of industry and research and is of prime interest to the electro-chemist. In simple terms, it is a measurement of the effective acidity or alkalinity of a solution. An acid or alkali, when dissolved in water, conducts electricity because of the occurrence of dissociation into particles or ions carrying a negative or positive charge and moving freely in the solution. It is the ratio of positive to negative ions that conveys the pH value. This is described in a scale extending from 0 to 14, on which pH7 represents a neutral value (pure water), decreasing numbers (7–0) indicate increasing acidity and increasing numbers (7–14) indicate increasing alkalinity.

A well-known method of measurement is that shown in Figure 2.20. This is based on the fact that a suitable glass membrane, M, separating two solutions, A and B, develops an e.m.f. across it which is directly related to the difference in hydrogen ion concentration of the two solutions. One of the two solutions is a stable pH liquid (usually pH7) so that the measurement is a relative one. In order to preserve a high impedence between the electrodes, an operational amplifier having a differential input conveys the difference e.m.f. to the recording or measuring equipment.

2.9.3   VACUUM MEASUREMENT

Continuous monitoring of vacuum pressure is required for many situations in nuclear physics. Most of the methods used fall into three categories, dependent on thermal conductivity, ionisation or range of radioactive particles.

A simple thermocouple method suitable for the range between 10 and 10^{-3} mm of mercury (Hg) is to associate a thin wire to the sensitive junction of a thermocouple. The wire is heated by passing a constant current through it. The heat dissipation, measured by the thermocouple, depends on the density, pressure and specific heat of the surrounding gas. All variables are kept constant except for

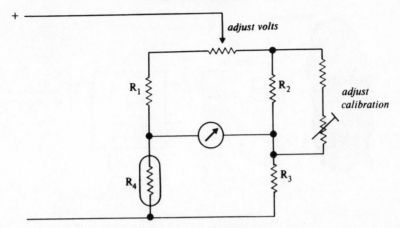

Fig. 2.21   A Pirani gauge for vacuum measurement

pressure, i.e. state of the vacuum. This is found to be simply related to heat dissipation and hence to the e.m.f. generated at the thermocouple junction.

The **Pirani gauge** is a similar device, but here the temperature of the heated wire is measured by including it in a Wheatstone bridge circuit and determining its resistance by off-balance of the bridge.

In order to measure very low vacuum pressures in the range $10^{-4}-10^{-8}$ mm Hg, the **ionisation gauge** is used. This is a thermionic triode valve having a directly heated filament and is sealed into the vacuum system (Figure 2.21). The grid of the triode is made > 100 V positive and the anode some 20 V negative. Electrons reaching the grid from the filament ionise the gas and ions are formed between the grid and anode. Some of these ions are drawn to the anode and the current which they carry can be made proportional to the pressure of the gas. It is essential to keep the electron current stable by adjustment to the filament current, and emission-stabilised circuits form an important element in the design of the gauge.

Measurement of pressure through radiation is a useful method which has some operational advantages, although it is not so convenient for providing a continuous signal of vacuum level. The device used is known as an **alphatron** and may be used to make a wide range of pressure measurement between 10^{-3} and 10^3 mm Hg.

The range of alpha particles in air has a characteristic shape if the alpha particles are monoenergetic. Instead of varying the distance between source and detector, a similar effect is obtained by varying the density of gas and hence pressure by reference to standard range curves.

2.10   Signal Conditioning

It will have become apparent from the preceding discussion that transducers need a given electrical environment in order to function correctly. Passive

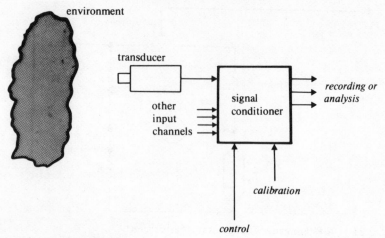

Fig. 2.22   Signal conditioning

transducers, such as the resistance strain gauge, require a stable constant current or voltage supply and additional resistors to complete a null bridge arrangement. Some energy-conversion types of transducer have to be coupled to a high-input impedance amplifier if their sensitivity is not to be seriously reduced. Additionally, many transducers require calibration before use and the necessary calibrating components will have to be made available.

Signal conditioning may be considered as comprising all those operations which are ancillary to the functioning of the transducer itself, and which are necessary in order to extract a signal from it related to the physical quantity being measured. A signal conditioner is the name given to the unit which carries out the process of signal conditioning between the transducer and the recording or analysis equipment (Figure 2.22).

Two essential features of a signal conditioner are a stable operational amplifier having a defined gain, high input and low output impedances, and a stable constant voltage (or current) power supply. The function and use of the operational amplifier, which is the name given to a highly stable and accurate amplifying device, are described in some detail in the next chapter. It may also be used as a basis for a frequency-selective filter or a charge amplifier, both of which may be required in the signal conditioner [13, 14].

Where piezo-electric transducers are used, the voltage amplifier is replaced by a charge amplifier which permits a reduction in the effects of cable capacitance on the transducer gain and frequency response [15].

A low-pass filter may be required in order to reject the high-frequency components generated by the transient excitation of an accelerometer, or to reduce the noise input from transducers situated at the end of long transmission lines.

Certain types of transducers, including variable reluctance and other transducers, will require an energising alternating current source. This will be modulated by the action of the transducer, so that detection of the modulated signal will need to be carried out in the signal conditioner. This is arranged by including a phase-sensitive detector, followed by a low-pass filter in order to remove the carrier and intermodulation harmonics. Finally, calibration resistors and the bridge-completing and balancing resistors of a strain-gauge transducer, also form elements in the signal conditioning process.

It will be seen from this that the signal conditioner performs many different functions and that these will differ according to the type of signal being measured and the transducers used. For this reason, the signal conditioner unit is designed in modular form to incorporate any number of modules required for a particular application.

A basic frame system available from several manufacturers consists of a wired series of sockets and integral power supplies, switches, metering, etc. The sockets connect with individual modules which may be inserted into slots in the unit frame system and make connection with other units and power supplies through the unit back wiring.

The modules carry their individual control adjustments (gain, calibration, range factor, etc.) and are designed to perform one or more of the functions described previously. Multi-channel operation is often required and several channels may be provided with each unit module. In larger units of this type, the role of the signal conditioner may expand to include signal operations of the pre-processing type, as discussed in the following chapter.

References

1. NEUBERT, H. K. P. *Instrument Transducers*. Oxford University Press, Oxford, 1975.
2. WOOLNET, G. A. *Transducers in Digital Systems*. Peter Peregrinus, London, 1977.
3. NORTON, H. N. *Handbook of Transducers for Electronic Measurement*. Prentice-Hall, Englewood Cliffs (NJ, USA), 1969.
4. OLIVER, F. J. *Practical Instrumentation Transducers*. Hayden Book Company, New York, 1971.
5. SYDENHAM, P. H. *Transducers in Measurement and Control*. University of New England Publishing Unit, Armidale, Australia, 1975.
6. ZIAS, R. H. and HARE, W. F. J. Integration brings a generation of low-cost transducers. *Electronics*, 45 1972, 83–88.
7. HARRIS, C. M. and CREDE, C. E. *Shock and Vibration Handbook*, (Vol. 1). McGraw-Hill, New York, 1961.
8. SPITZE, F. and HOWARD, B. *Principles of Modern Instrumentation*. Holt, Rinehart & Winston Inc., New York, 1972.
9. COLLETT, C. V. and HOPE, A. D. *Engineering Measurements*. Pitman, London, 1974.
10. BENNETT, E. J. Accurate measurement of flow by turbine flowmeters. *Measurement and Control*, 12, 1979, 59–54.
11. SPINK, L. K. *Principles and Practice of Flow Meter Engineering*. Plimton Press, Norwood, Mass., 1967.

12.   NEWMAN, D. W. *Instrumental Methods of Experimental Biology*. Macmillan, London, 1963.
13.   GRAEME, J. G. *Application of Operational Amplifiers* (Ch. 3). McGraw-Hill, New York, 1973.
14.   EARLEY, B. *Practical Instrumentation Handbook*. Scientific Era Publishers, Stamford, 1976.
15.   WIGHTMAN, E. J. *Instrumentation in Process Control*. Butterworth, London, 1972.

ADDITIONAL REFERENCE

DOEBLIN, E. O. *Measurement Systems; Application and Design*. McGraw-Hill, New York, 1966.

Chapter 3

Pre-processing

3.1 Introduction

In this chapter we will consider signal amplification and frequency filtering — two operations which are likely to be carried out on any signal which we wish to record. Indeed, we may find it necessary to carry out other forms of signal modification as well, such as trend removal, decimation and calibration before we are in a position to display or record the signal for analysis. We call these operations, carried out prior to analysis, **pre-processing** and they are nearly always carried out at the time of data acquisition.

Signal information is almost never obtained in precisely the right form to suit the analysis methods available. We find this situation, for example, with analog signals where pre-processing is required prior to recording, analysis or conversion to digital form, and also with digital data which often require modification before they can be accepted as input to a digital analysis program.

The need for pre-processing may be seen if we consider the practical difficulties inherent in, for example, the 'capturing' of a transient effect occurring at an imprecise time but within a given time interval. This time interval may be long and, as a result, the useful record period may be unduly extended. Apart from the errors that this will introduce in the analysis estimates obtained from processing the summation of background noise plus signal, the analysis process will become unnecessarily protracted and hence become more expensive. This is particularly the case for digitised data, where the analysis methods are likely to be time-consuming, whether applied to the background or to the transient signal. Therefore, arrangements need to be made to delay the recorded transient, using control or elimination methods, and to terminate analysis following the completion of the transient so that only the transient itself will be analysed.

The recorded signal may itself represent a mixture of desired signal plus other information and we may wish to carry out a separation process. For example, if the required signal contains noise or an unwanted discrete frequency component, then its time history can be represented as

$$x(t) = s(t) + n(t) \tag{3.1}$$

where $s(t)$ represents the desired signal and $n(t)$ represents the noise signal. Equation (3.1) may be transformed to give a similar equation in the frequency domain, i.e.

$$X(\omega) = S(\omega) + N(\omega) \qquad (3.2)$$

If $S(\omega)$ is sufficiently remote from $N(\omega)$ along the frequency spectrum, it is possible to remove this by filtering in the frequency domain and to carry out an inverse transform to recover the value $x(t) = s(t)$, which represents the wanted signal.

Filtering is indeed an important element in pre-processing operations. As will be shown in Chapter 5, continuous or analog filtering is necessary to precede digitisation if the distorting effects of aliasing are to be avoided. Some processing of the digitised data will also be required and discrete or digital filters have been developed for this purpose. Both continuous and discrete filters will be described in this chapter.

Other constraints necessary in order to extract meaningful data from the raw signal may be found in the amplitude or power domain. We may, for instance, wish to reject the signal if its standard deviation falls outside a permitted value. This implies continuous monitoring of the raw signal to detect this characteristic, and a control action to modify the processing operation. Sometimes the unwanted information added to the required signal can represent a modulation of the signal by the characteristics of the measuring device, such as a decrease in sensitivity as a function of temperature or time. The removal of this function is often termed **trend removal**.

We may also find that the refined data for analysis is defined or represented by a combination of signals recorded on a number of separate channels. To achieve a signal-to-noise enhancement, ensemble averaging by parallel addition of these signals may be all that is required. Other more complex techniques may be necessary before the signal is sufficiently clear for useful analysis.

An associated processing task of some importance is the calibration and identification of the acquired data. The raw signal or data consist essentially of a sequence (continuous or discrete) of numbers representing a measured parameter (e.g. pressure, displacement, temperature, velocity, etc.). In order to carry out calculations on these numbers, conversion into the appropriate units is needed, requiring a knowledge of the conversion factor, i.e. calibration is required. This may form part of the recorded ensemble and be given as a separately recorded sequence, or it may emerge as a result of earlier measurement. The process of calibration may be obtained within the computer analysis program, but it is often convenient to carry out the necessary arithmetic operations in the pre-processing stage in order to improve accuracy or dynamic range of the signal.

Finally, the pre-processing operations will almost certainly include various organisational or 'labelling' tasks necessary so that the signals or data can be properly identified as belonging to a particular processing operation. This is essential, not only to avoid errors but also to reduce handling time following data acquisition. We can carry this out by including both run and task identification during the recording or pre-processing operation. Where analog recording is employed, a separate channel can be coded to indicate this essential housekeeping information. The first few blocks of a digital tape can be similarly allocated. This information can be accessed later and included with the processed output signal, to ensure that identification information is always associated with the data.

3.2. Signal Amplification

The most important element in a pre-processing system is the amplifier situated between the source transducer and the processing or recording system. This amplifying element may not, strictly speaking, amplify at all but simply provide an impedance-matching function. However, it performs the initial conditioning of the acquired signal and the accuracy of the complete system will be dependent upon it.

The basic constituent of a signal amplifier is the **operational amplifier**. This is a high-gain, solid-state device which is capable of operation down to zero frequency (d.c. level). When used in conjunction with simple feedback circuits it is capable of providing a moderate gain with high stability and a performance determined almost entirely by the values of the resistive feedback elements. We shall not be concerned here with the design of the operational amplifier itself but with its use as a circuit component in signal amplification systems [1–5].

The modern operational amplifier is a small encapsulated unit making use of integrated circuit technology in which the active elements are field-effect transistors (FET) or metal oxide silicon (MOS) bipolar transistors. To function adequately, it must have a very high gain (at least 10^5) and be carefully designed for stability and freedom from d.c. voltage drift. Apart from its power supply connections, it will usually have two input terminals in addition to a common (earth) terminal so that balanced input can be applied. The output is unbalanced, i.e. one terminal only, and at a low impedance level, while the input has high impedance so that only a negligibly small input current is drawn from the signal source. A simple signal amplifier providing unit gain and sign inversion is shown in Figure 3.1. The output voltage is $-A\,e_b$, where A is the amplification or gain factor and e_b is the voltage at the true input to the amplifier, which will be different from e_i, the voltage at the input terminal. For unit gain, $R_i = R_f$. The association of the input resistor, R_i, and the feedback resistor, R_f, with the

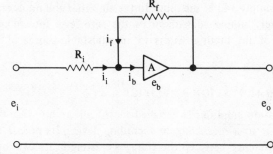

Fig. 3.1   Signal amplification using an operational amplifier

operational amplifier gives a large measure of negative feedback to the amplifier circuit and this has a number of advantages. To understand these we will first look into the characteristics of this feedback circuit.

Referring to Figure 3.1, the currents at the true input to the operational amplifier shown as x, may be summed using Kirchhoff's current law thus

$$i_i = i_b + i_f \tag{3.3}$$

Since i_b is negligibly small, we may put $i_b = 0$ and write

$$i_i = \frac{e_i - e_b}{R_i} = \frac{e_b - e_o}{R_f} = i_f \tag{3.4}$$

Replacing e_b by $-e_o/A$, where A is the gain of the operational amplifier

$$\frac{e_i + e_o/A}{R_i} = \frac{-e_o/A - e_o}{R_f} \tag{3.5}$$

and the voltage gain with feedback is

$$\frac{e_o}{e_i} = \frac{-AR_f}{R_f + R_i + AR_i} \tag{3.6}$$

which may be rearranged as

$$\frac{e_o}{e_i} = -\frac{R_f}{R_i}\left[\frac{1}{1 + (1 + R_f/R_i)/A}\right] \tag{3.7}$$

Since A is very large, typically $> 10^5$, then, to a very close approximation, equation (3.7) becomes

$$\frac{e_o}{e_i} \simeq -R_f/R_i \tag{3.8}$$

Thus the gain of the signal amplifier is essentially independent of operational amplifier characteristics and determined entirely by the value and precision of R_i and R_f.

There is another advantage of a voltage null at the true input to the amplifier. If several input signals are to be mixed, i.e. the amplifier is to be used as a summing device, then the input circuits are quite independent of one another. The input currents simply add at the input to the amplifier and no interaction occurs. This permits very simple adjustment of the zero level for an input signal by returning one of the input resistors to an adjustable source of d.c. potential (Fig. 3.2).

3.2.1 INSTRUMENTATION AMPLIFIERS

An **instrumentation amplifier** is designed specifically to interface transducer and other equipment to a measuring or recording device. Its precise function is to measure the difference between the voltages existing at its two input terminals,

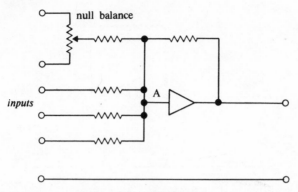

Fig. 3.2   A summing amplifier with null balance

to amplify this difference by a precisely set gain, and to present the result between a pair of terminals at the output circuit. The specification for such an amplifier, which will include one or more operational amplifiers as individual components, will include a balanced input, a high input impedance, low offset and drift, low non-linearity, stable gain and low effective output impedance.

We have seen how the characteristics of the operational amplifier will enable a number of these specifications to be realised. To achieve a balanced input or **differential amplification** some circuit elaboration is necessary.

A balanced input is necessary for many reasons. The transducer system may be located some considerable distance from the pre-processing unit and the long electrical leads will be susceptible to induced signals from adjacent power lines or electrical equipment and from the existence of long earth loops. Considerable improvement in the performance of such systems is obtained when balanced line connections are used to an operational amplifier designed to have a high **common mode rejection**, namely the rejection of signals established between each line and a common earth in favour of signals between the two lines. These are known as **differential amplifiers** and one version is shown schematically in Figure 3.3. Assuming ideal amplifier performance and precision of resistors values the circuit performance is defined by

$$e_o = (e_2 - e_1)\frac{R_f}{R_i}$$

$$R_{input} = 2R_i$$

$$R_{common} = \frac{1}{2}\left[R_i + \frac{R_f \cdot R_{cm}}{R_f + R_{cm}}\right] \tag{3.9}$$

where R_{cm} is the common mode input resistance of the operational amplifier itself.

A number of difficulties arise with this circuit. The input resistance is dependent on R_i and, if this is made large to reduce loading on the transducer, then increased offset and drift will occur and possibly an increase in noise level as well. The common mode rejection is highly dependent on R_i and R_f and very close tolerances on value are demanded.

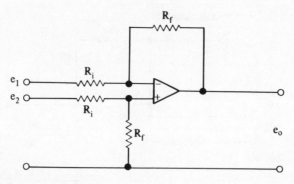

Fig. 3.3   The differential amplifier

If the resistors are not perfectly matched then a common mode input signal will give rise to a differential signal at the amplifier input terminals which will appear as an amplified signal at the output. Thus, the common mode rejection capability of the circuit of Figure 3.3 will always fall below that of the amplifier used in the circuit.

Various methods of overcoming these problems are available and one such solution is shown in Figure 3.4. This uses two non-inverting amplifiers with the differential inputs connected directly to one input of each amplifier. The gain of this circuit is given as

$$e_o = (e_2 - e_1) \cdot \left[1 + \frac{R_f}{R_i} \right] \tag{3.10}$$

and is theoretically zero if $e_2 = e_1$, although the common mode rejection is again critically dependent upon resistor matching.

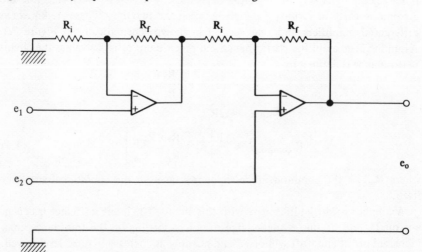

Fig. 3.4   An alternative differential amplifier

Voltage offset and offset drift in instrumentation amplifiers refer to the departure from zero level output value, initially and as a function of time, for a zero input differential voltage. The value of this offset is a function of the gain of the amplifier. The offset, measured at the output, is equal to a constant plus a term proportional to gain. Specifications of instrumentation amplifiers usually refer the output offset value to the input at a particular gain value, with the off-set constant essentially the offset at unity gain. Offset drift is a function of temperature and is also referred to the input value at a specific gain and given as so many $\mu V/^\circ C$ for that gain. Manufacturers' data sheets usually provide curves of offset voltage against gain and temperature.

A specification value of some importance for transient amplification is that of **settling time**. Settling time is defined as that length of time required for the output voltage to reach and maintain its final value within a certain tolerance, usually 0.1%. It is usually specified for a fast step-input voltage that will drive the output through its full-scale range. It is important to note that settling time is not necessarily proportional to gain, so that a reduction in amplifier gain will not always reduce the settling time by the same value.

3.2.2 CAPACITATIVE AMPLIFIERS

The operational amplifier is a versatile active element and is found in many analog signal processing systems. We have considered it here as a component of an instrumentation amplifier for balanced or unbalanced inputs. Later in this chapter, it will be seen to form a basis for continuous active filters. It also has a role to play as an integrating element in which the feedback resistor, R_f of Figure 3.1, is replaced by a capacitor. We know that the current flowing through R_i must be $(e_i - e_b)/R_i$ and, as i_b is negligibly small, then this current must flow into the capacitor, C_f, which now takes the place of R_f. The charge that accumulates in C_f must be the integral of this current. At the same time, we know that the voltage difference across C_f must be the charge it contains divided by its capacitance. Thus we have

$$(e_b - e_o)C_f = \int \frac{(e_i - e_b)}{R_i} \, dt$$

but

$$e_o = -Ae_b$$

and we can write

$$e_o + \frac{e_o}{A} = -\int \frac{e_i}{R_i C_f} \, dt \simeq e_o \tag{3.11}$$

Thus, e_o is simply the integral of e_i times a constant, determined by a suitable choice of R_i and C_f.

This is the basis for the integrating and differentiating elements of an analog computer and for a number of specialised amplifying circuits for pre-processing and signal conditioning. Two illustrations of the capacitative amplifier will be given. The first is the **charge-compensating amplifier** and the second is the **sample and hold amplifier**.

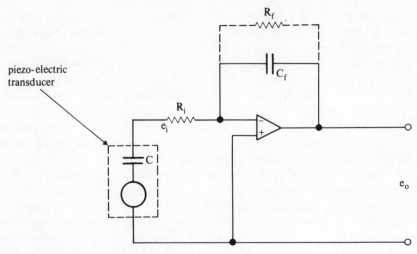

Fig. 3.5   Amplifier for a piezo-electric device

3.2.3   THE CHARGE-COMPENSATING AMPLIFIER

This is an application of the operational amplifier to the special problems of amplifying the output from a piezo-electric transducer. The piezo-electric transducer is essentially a capacitative device producing a charge proportional to the physical variable under investigation (see Ch. 2). Charge can be computed by adding, or rather integrating, the current over a period of time. Measurement of charge is carried out by the operational amplifier current integrator shown in Figure 3.5. A feedback capacitor is used in place of R_f and produces an output voltage

$$e_o = -\frac{KmC}{C_f} \qquad (3.12)$$

where K is a constant associated with the particular type of transducer and m is the measurement variable. From this equation, we see that the effects of the transducer capacitance are nullified by the integrating action. It is necessesary in a practical amplifier configuration to shunt C_f by a feedback resistor, R_f, allowing the charge to leak away. This has the effect of limiting the amplifier bandwidth to a frequency

$$f = \frac{1}{2\pi C_f R_f} \text{ Hz} \qquad (3.13)$$

which is not usually serious.

3.2.4.   THE SAMPLE-AND-HOLD AMPLIFIER

We will meet this application again in Chapter 5, where it will be discussed as a device used for maintaining the input to the analog-to-digital converter constant at a value representing the analog input occurring at a certain precisely known time. A design for a sample-and-hold amplifier (SHA) is shown in Figure 3.6b. This consists of a storage capacitor, two operational amplifiers and a switch actuated by the sampling clock pulse generator.

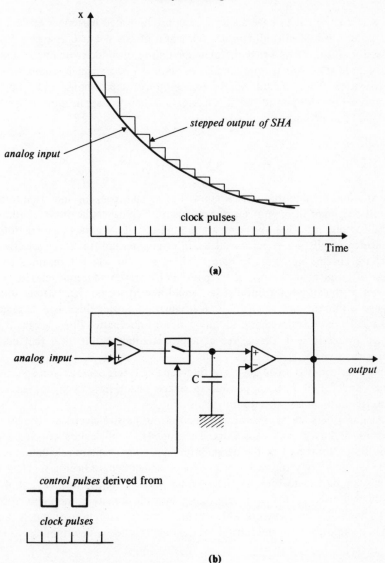

Fig. 3.6   A sample and hold amplifier

During the *sample* phase, the switch is closed and the capacitor is included in a high-gain feedback loop which rapidly charges the capacitor to the potential value existing at the input terminals. During the *hold* phase the switch is open and the capacitor is disconnected from its charging source and, ideally, retains its charge which is conveyed through the unity-gain second amplifier to present a constant output potential. In the sample mode, the performance of the SHA can be characterised by specifications similar to those of the operational amplifier, namely offset and drift, non-linearity, gain, and input—output impedances.

We would expect it to have a slower response indicated by longer settling time and poorer gain—bandwidth product because of the need to charge a storage capacitor. Also, we find that the dynamic nature of mode-switching to and out of the hold state will cause a number of errors to occur which require accurate specification. These include aperture uncertainty, step response and droop and are more properly considered in Chapter 5, when the application of the SHA to digital conversion is described.

3.3   Filters

Linear filters are important in the context of signal processing and often form an essential element in pre-processing equipment. Typical applications are found in signal-to-noise ratio improvement, smoothing of data, bandwidth reduction and avoidance of aliasing effects. Whilst a filter implies any frequency-selective device, we confine our remarks here to those systems which transmit a certain range of frequencies, known as the pass-band, or reject all frequencies in a range known as the stop-band. Ideally, we would like to see no signal attenuation or change within the pass-band and total attenuation outside this range. In practice, this ideal is not achieved and the discussion of alternative filter designs is concerned primarily with the degree of approximation to this ideal that may be achieved.

We consider two broad classes of filter; analog filters operating on continuous signals and digital filters applicable to signals which have been quantised and converted into binary digital form.

Analog filters are designed traditionally from passive electrical elements, i.e. resistors, inductors and capacitors. The impedance of devices associated with such filters forms part of the design information and element calculation can be quite complex. In more recent years, these elements have been associated with amplifying devices to form an 'active' filter which is almost independent of load impedance effects. Of major importance in such designs is the elimination of bulky inductive components which permits the construction of much smaller filters using capacitors and resistors as the only frequency-determining elements.

3.3.1   ANALOG ACTIVE FILTERS

Whilst the detailed design of analog filters falls outside the scope of this book [6—8], it will be necessary to review some of the concepts used in such designs and to consider the performance of certain well-known types of analog filter.

The operation of an analog filter (or any other linear electrical system) may be considered in terms of its **transfer function**. We define the transfer function $H(\cdot)$ of a linear system as the ratio of the transformed input and output functions. The most general way to consider this transformation is through the use of the Laplace transform. This approach also enables the signal itself to be represented by a set of s-plane poles and zeros and is much used in filter design.

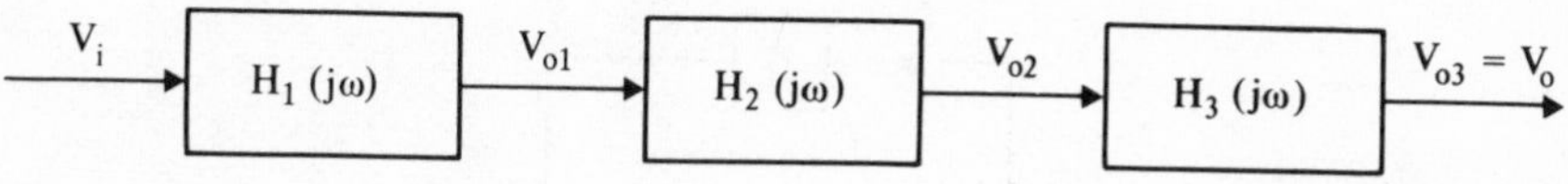

Fig. 3.7   A series of cascaded filters

However, we may, as a general rule, derive a sinusoidal frequency spectrum by substitution of the complex frequency $j\omega$ for the Laplace variable s, and represent the transfer function in terms of the Fourier transform. Since our primary interest is in the operation rather than the design of digital filters we have taken this approach here and represent the transfer function by $H(j\omega)$, where $\omega = 2\pi f$.

Using the transfer function as a complex output:input ratio in the frequency domain, we can describe the operation of a physical system in an easily manipulative form. Even the most complicated system can be treated as a four-terminal 'black box' having a two-port entry and exit route.

Designating V_i and V_o as the input and output signals, respectively, we may write for the transfer function

$$H(j\omega) = \frac{V_o(j\omega)}{V_i(j\omega)} \tag{3.14}$$

Transfer function theory may be applied to the serial linear system shown in Figure 3.7. This consists of a number of individual systems, in this case a series of cascaded filters, where the overall transfer function $H(j\omega)$ becomes the product of the individual transfer functions $H_1(j\omega)$, $H_2(j\omega)$ and $H_3(j\omega)$. Thus

$$\frac{V_{o_1}(j\omega)}{V_i(j\omega)} = H_1(j\omega)$$

$$\frac{V_{o_2}(j\omega)}{V_{o_1}(j\omega)} = H_2(j\omega)$$

$$\frac{V_o(j\omega)}{V_{o_2}(j\omega)} = H_3(j\omega) \tag{3.15}$$

and

$$\frac{V_o(j\omega)}{V_i(j\omega)} = \frac{V_{o_1}(j\omega) \cdot V_{o_2}(j\omega) \cdot V_o(j\omega)}{V_i(j\omega) \cdot V_{o_1}(j\omega) \cdot V_{o_2}(j\omega)} = H(j\omega) \tag{3.16}$$

This is an important feature of transfer function manipulation and finds wide use in the determination of a complete system's characteristic in terms of its subsystems and elements.

A simple form of active filter is shown in Figure 3.8a. This consists of a high-gain operational amplifier associated with two passive networks described by the single impedances Z_i and Z_f. The amplifier is assumed to have a high-level response down to zero frequency and to provide an open-loop gain of the order 10^5–10^8. It was shown earlier (cf. equation 3.8) that the transfer function of this system may be represented with very little loss of accuracy as

$$H(j\omega) = -Z_f/Z_i \tag{3.17}$$

where the negative sign implies signal inversion by the amplifier.

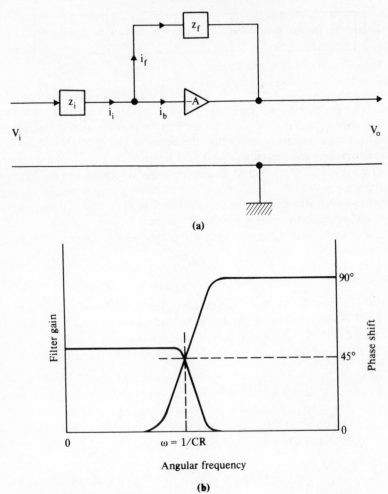

Fig. 3.8(a)   An active filter
        (b)   Phase–amplitude response

As a consequence of the very high internal gain of the amplifier, the difference current $i_b = i_i - i_f$ is extremely small, so that terminal B (the true input to the amplifier) is practically at ground potential. Thus Z_i and Z_f become the short-circuit transfer impedances

$$Z_i = V_{i/i_i} \quad \text{and} \quad Z_f = V_{o/i_f} \qquad (3.18)$$

This permits the numerator and denominator of the transfer function to be treated separately and considerably simplifies the determination of the complex impedance characteristic required in a given transfer function.

In order to illustrate this type of approach, we consider first a filter having the transfer function

$$H(j\omega) = \frac{A}{1 + Tj\omega} \qquad (3.19)$$

This represents a first-order lag circuit, where $A = R$ and $T = CR$. The modulus of $H(j\omega)$ gives the frequency response and the ratio of the real and imaginary parts of the phase angle of lag Φ, viz.

$$|H(j\omega)| = V_o/V_i = \frac{1}{\sqrt{[1 + (CR\omega)^2]}} \qquad (3.20)$$

and

$$\Phi = \cotan(-CR\omega) \qquad (3.21)$$

The phase angle is 0 for $\omega = 0$ and -0 as $\omega \to \infty$. The amplitude ratio may be rewritten in decibel notation as

$$20 \log_{10}|G(j\omega)| = 20 \log_{10} \cdot \sqrt{[1 + (CR\omega)^2]} \qquad (3.22)$$

When $CR\omega \ll 1$, then the gain characteristic $\simeq 20 \log_{10} 1 = 0$, which represents a level or zero db slope, and when $CR\omega \gg 1$ then

$$-20 \log_{10} \sqrt{1 + (CR\omega)^2} \simeq -20 \log_{10}(CR\omega)$$

which represents a slope of -6db per octave frequency change.

A phase–amplitude plot for this filter is given in Figure 3.8b where it will be seen to constitute a low-pass filter having the performance characteristics described above.

It is normal practice to use the low-pass filter as a basis for developing filter approximation functions. Further, it is advantageous to arrange the filter network transfer function to correspond to certain mathematical functions, for example the Butterworth and Chebychev polynomials, since these provide filter amplitude and phase characteristics which are optimum in one sense or another for specific applications. Definitions of filter performance for Butterworth and Chebychev filters are more usefully expressed in terms of the squared-magnitude transfer functions

$$|H(j\omega)|^2 = \frac{1}{1 + (\omega/\omega_c)^{2n}} \qquad \text{(Butterworth) (3.23)}$$

and

$$|H(j\omega)|^2 = \frac{1}{1 + E^2 C_n^2(\omega/\omega_c)} \qquad \text{(Chebychev) (3.24)}$$

where ω_c is the filter cut-off frequency, C_n is a Chebychev polynomial of order n and E is a constant.

The Butterworth low-pass filter is characterised by the property that its amplitude response is maximally flat in the vicinity of zero frequency and is hence optimum in this sense. The response of the filter is completely specified by the order n of the filter, where n is the number of filter poles. The attenuation performance approaches that of an ideal filter as n becomes very large. However, its transient response deteriorates with n.

Chebychev filters provide equi-ripple pass-band behaviour and a monotonic stop-band response. They are also optimal but in a different sense, providing the best possible compromise considering both pass-band and stop-band amplitude

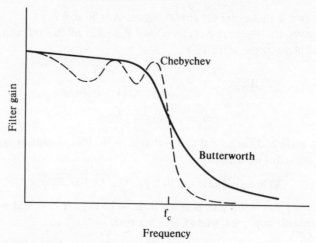

Fig. 3.9   Butterworth–Chebychev filter performance

response. A sharper response is obtained for a given value of n compared with the Butterworth filter, but at the expense of an oscillatory transient response.

Comparative amplitude responses for these two filters are shown in Figure 3.9. These filters represent particular approximations to the ideal filter characteristics. Other approximations, based on different polynomials, are also used. One of these is the Bessel filter which approximates an ideal phase characteristic, namely constant transmission delay, but with poorer attenuation response [9, 10].

A widely used active filter is the second-order filter, with $n = 2$, shown in Figure 3.10. This is known as a unity-gain network filter, having the two forms shown in Figure 3.10 for low-pass and high-pass operation. Excellent performance is obtained in terms of thermal drift and component tolerance, and the impedance characteristics are such that a number of such filters may be cascaded to provide higher-order filtering.

Tables of component values for the RC elements are available for various filter characteristics (Butterworth, Chebychev, etc.) and cut-off frequencies [6, 7], so that it is a simple matter to construct a filter using a unity-gain operational amplifier as the active element.

3.3.2 SOFTWARE DIGITAL FILTERS

A digital filter may be realised either by suitably interconnecting electronic logic elements or by programming a digital computer. We shall consider this latter method first.

A digital signal consists of a series of sampled and quantised values forming a series, x_i, where $i = 1, 2, \ldots, N$. Performing a filtering action on this series means, in practice, that the individual members of the series are multiplied and combined in such a way as to produce a new modified series, y_i, which when converted into a continuous signal will have had its frequency characteristics modified

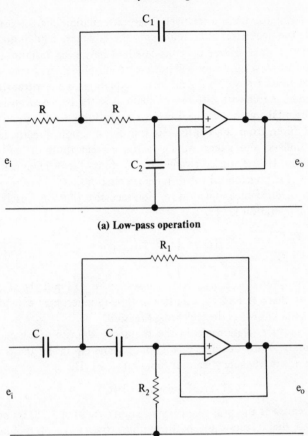

(a) Low-pass operation

(b) High-pass operation

Fig. 3.10　Unity gain network filter
(a) Low-pass
(b) High-pass

in the desired way. We represent this mathematically by means of a linear difference equation

$$y_i = \sum_{k=0}^{P} b_k y_{i-k} + \sum_{k=0}^{M} a_k x_{i-k} \ (i = 1, 2, \ldots, N) \qquad (3.27)$$

where P and M are positive integers and a_k and b_k are real constants. When M = 0 then the filter is auto-regressive and implies a form of feedback analogous to the analog active filter. This is known as the **recursive filter**. For P = 0 it reduces to a non-recursive form, which is analogous to the analog passive filter and is hence inherently stable.

The two classes of filter have quite different properties. The **non-recursive filter** represents the summation of a limited number of input terms and thus has a finite memory. It can have excellent phase characteristics but requires a large number of terms to obtain a relatively sharp attenuation characteristic. On the other hand, the recursive filter represents the summation of both input and output terms so that it can be considered as having an infinite memory. It requires relatively few terms but will possess poorer phase performance.

The appropriate transform for difference equations is the **z-transform** which performs the same role with difference equations as the Laplace transform carries out for differential equations. The use of the z-transform permits the specification of a digital filter from the continuous transfer function directly in terms of delays, multipliers, and adders which is the correct form of hardware filter implementation. Its use for 'software' filters permits the development of a convenient form of expression for the difference equations.

The form of difference equation for non-recursive filters is closely related to a convolution operation [11].

$$y(t) = \int_{-\infty}^{\infty} h(\tau)x(t-\tau)d\tau \qquad (3.28)$$

This gives the filtered response, $y(t)$ of a system to an input signal, $x(t)$ where the system is characterised by a particular time-function $h(\tau)$ extending over a given period and known as the **weighting function**.

There is another relationship in the frequency domain which expresses the filtered response $Y(\omega)$ in terms of the input signal $X(\omega)$ and the transform of $h(\tau)$, known as the frequency transfer function $H(\omega)$. This is

$$Y(\omega) = X(\omega) \cdot H(\omega) \qquad (3.29)$$

Now if we represent the process of convolution (equation 3.28) by means of an asterisk, then the relationships between these two ways of looking at non-recursive filtering can be summarised as

$$
\begin{array}{c}
y(t) = x(t)*h(\tau) \\
\uparrow \qquad \quad \downarrow \text{DFT} \quad \downarrow \text{DFT} \\
\text{IDFT} \quad X(\omega) \cdot H(\omega) = Y(\omega)
\end{array}
\qquad (3.30)
$$

where DFT and IDFT represent the direct and indirect Fourier transform, respectively. Thus, if we take the product of the Fourier transforms for the signal, $x(t)$ and the filter weights, $h(\tau)$ namely, $X(\omega) \cdot H(\omega)$, we obtain a filtered response in the frequency domain, $Y(\omega)$ which when subject to an **inverse** Fourier transformation gives exactly the same filtered response in the time domain, $y(t)$ that we would obtain by the convolved product of $x(t)$ and $h(\tau)$. This inter-relationship is important, since it provides a technique for filtering via the Fourier transform which is more economical in computing time than the direct convolution method.

To summarise the process of digital filtering we can now recognise three fundamental techniques. These are:

(i)   convolution (direct filtering)
(ii) Fourier transformation (indirect filtering)
(iii) auto-regression (use of difference equations).

The first two techniques are generally confined to the non-recursive filter and the third to the recursive filter.

The recursive filter is particularly suitable for implementation on small computers, such as the microprocessor, and may form part of the pre-processing requirements for digital data. A simple example of a first-order difference equation for a low-pass recursive filter is given by

$$y_i = x_i + Ky_{i-1} \ (i = 1, 2, \ldots, N) \tag{3.31}$$

The transfer function can be obtained by letting $y_i = A \sin i\omega t$, where $\omega = 2\pi f$ and $f = 1/2T$. Thus

$$y_i = KA \sin i\omega(t - T) + x_i$$

$$= KA \sin i\omega t \cdot \cos i\omega T - KA \cos i\omega T \cdot \sin i\omega T + x_i$$

But $\sin i\omega T = \sin i(2\pi T/2T) = \sin i\pi = 0$, since i is an integer. Therefore $y_i = Ky_i \cos i\omega T + x_i$ and

$$H(t) = y_i/x_i = \frac{1}{1 - (K \cos i\omega T)} \tag{3.32}$$

K can be related to filter cut-off frequency by defining f_c to be at the half-power point

$$|H(t)|^2 = \frac{1}{2} = \frac{1}{1 + (K^2 \cos^2 i\omega T) - (2K \cos i\omega T)}$$

but $\cos^2 i\omega T = \cos^2 [i(2\pi T/2T)] = \cos i\pi = 1$, since i is an integer, so that

$$K^2 - 2K \cos \omega_c T = 1 \tag{3.33}$$

where $\omega_c = i\omega$ and $K = \cos \omega_c T \pm \sqrt{(\cos^2 \omega_c T + 1)}$. This can be substituted in the recursive expression given in equation (3.31) to enable a simple algorithm to be implemented for the low-pass filter. Note that only one multiplication and one addition is required to realise a single output filter point. To program this for the digital computer, we simply take in turn each value of the x_i series and add to it the value of the *preceding* value of y_i multiplied by the value of K, calculated as a subroutine from the filter cut-off frequency and data sampling interval. This sum forms a new filtered series value, y_i. For the first value of this series we need to give unit value to y_{i-1}.

Many design techniques for recursive digital filters are available and it is not possible to treat these fully here. One technique which forms the basis of software and hardware digital filters is that of the **z-transform**. One way of defining the z-transform is from the Laplace transform and, since filter theory for continuous analog filters is generally expressed in Laplace form, we can consider the z-transform as a logical extension of this for a discrete series.

A sampled series, x_i, can be considered to be the product of a continuous signal, x(t), and a set of uniformly spaced unit impulses. Thus, the Laplace transform, H(s), of a sampled series can be expressed as

$$H(s) = \int_0^\infty [a_0\delta(t) + a_1\delta(t-T) + a_2\delta(t-2T) + ,\ldots,]\exp(-sT) \cdot ds$$
(3.34)

where a_0, a_1, a_2, ..., represent the sample amplitudes and T is the sampling interval. Hence

$$H(s) = \sum_{k=0}^\infty a_k \exp(-ksT)$$
(3.35)

This may be expressed in simplified terms to facilitate the algebraic manipulation by replacing exp (sT) by z, thus

$$\exp(sT) = z \qquad \text{or} \qquad \exp(-sT) = z^{-1}$$
(3.36)

and by replacing H(s) by H(z) so that

$$H(z) = \sum_{k=0}^\infty a_k z^{-1}$$
(3.37)

H(z) is thus by definition the z-transform of x_i and represents a power series in z^{-1} with coefficient a_k representing the amplitude of successive samples of x_i.

We can regard the multiplicative factor z^{-k} as representing simply a time delay of k sampling periods on the amplitude coefficient, a_k. Thus we may write

$$H(z) = z(x(n-k)T) = z^{-k} \cdot x(nT)$$
(3.38)

which expresses the delay property or **shifting theorem** of the z-transformation. Each delay by one sampling interval corresponds to a multiplication by z^{-1} in the z-domain. Thus $H(z) = z^{-2}$ corresponds to a sample taken with a delay of two sampling units, i.e. x_{i-2}. Similarly $H(z) = z^3$ corresponds to a forward shift in time by three sampling intervals, i.e. x_{i+3}. This is, of course, very easy to implement using the digital computer, since we just have to select the samples having the appropriate delay and add or subtract their values to derive a filtered output sample y_i.

An example of this method of design is the **bi-linear z-transform**. It is an indirect method, since the requirements of an equivalent analog filter are first postulated and then transformed into a discrete form from which the filter weights can be derived.

The analog filter is expressed as a Laplace transfer function, H(s), in which the integration operation, shown in equation (3.28), is given in terms of the s-operator as $1/s$. Essentially, therefore, the bi-linear z-transform means the translation of the integrating operation, $1/s$, into the z-transformation operation. This is carried out by means of a z-transfer function

$$H(z) = \frac{y(z)}{x(z)} = \frac{1}{2}T\left[\frac{1+z^{-1}}{1-z^{-1}}\right]$$
(3.39)

which is stated here without proof.

This represents a trapezoidal approximation to integration using the s-operator expressed as a function of z. Thus, the transfer function, H(s), for a continuous system can be replaced by the z-transfer function, H(z), using the relationships

$$\frac{1}{s} \rightarrow \frac{1}{2} T \left[\frac{z+1}{z-1} \right]$$

$$s \rightarrow \frac{2}{T} \left[\frac{z-1}{z+1} \right] \tag{3.40}$$

giving

$$H(z) \equiv H(s) \Big|_{s \,=\, \frac{2}{T} \left[\frac{z-1}{z+1} \right]} \tag{3.41}$$

This transform is known as the bi-linear z-transform. It maps the imaginary axis of the s-plane into a unit circle of the z-plane such that the left-hand side of the s-plane corresponds to the interior of the circle.

If the transformation is carried out precisely as a unique 1:1 relationship, the filter will be stable and of the same order as the original Laplace transfer function. The frequencies of the two relationships will, however, be different so that the replacement of s shown in equation (3.41) will result in a digital filter having a cut-off angular frequency ω_d, related to the continuous frequency ω_a, by

$$\omega_a = \frac{2}{T} \tan \left[\frac{\omega_d T}{2} \right] \tag{3.42}$$

This has the effect of compressing the complete continuous frequency characteristics into a limited digital filter frequency range of $0 < \omega T < \pi$, as indicated in Figure 3.11. In the practical application of equations (3.41) and (3.42), we will be calculating the ratio s/ω_a so that the $2/T$ term may be dropped and we can write

$$s = \frac{z-1}{z+1} \tag{3.43}$$

$$\omega_a = \tan \left[\tfrac{1}{2} \omega_d T \right] \tag{3.44}$$

The method of design using the bi-linear transform can be summarised as follows:

1. A new set of frequencies for the continuous transfer function, H(s), is calculated from the desired digital frequencies using equation (3.44).
2. A suitable continuous transfer function, H(s), is chosen to give the required filter performance using the frequencies derived from step (1).
3. The operator s in H(s) is replaced by a function in z and the new transfer function, H(z), expressed as a ratio of polynomials in z.
4. H(z) is converted into a difference equation by the application of the shifting theorem and the algebraic equation is rearranged to give a recursive equation for the output sample.

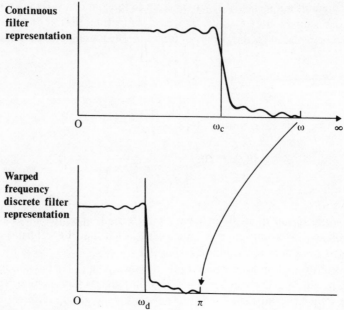

Fig. 3.11   Warping of the frequency scale

5. The recursive equation, considered as an algorithm, is programmed for the digital computer.

Let us look at this design applied to the low-pass Butterworth filter that we met earlier.

The continuous transfer function for a first-order ($n = 1$) filter may be derived in terms of the Laplace function as

$$H(s) = \frac{1}{1 + s/\omega_c} \tag{3.45}$$

A difference equation may be developed from this using the four equations (3.41) to (3.44). Given a cut-off frequency f_c in hertz (Hz) or cycles per second, the required equivalent continuous cut-off angular frequency is, from (3.44)

$$\omega_a = \tan(\pi f_c T) \tag{3.46}$$

The z-transform function is, from (3.43) and (3.45)

$$H(z) = \frac{Y(z)}{X(z)} = \frac{1}{1 + \left[\dfrac{z-1}{z+1} \cdot \dfrac{1}{\omega_a} \right]}$$

$$= \frac{z+1}{z\left[1 + \dfrac{1}{\omega_a} \right] + \left[1 - \dfrac{1}{\omega_a} \right]} \tag{3.47}$$

Multiplying by z^{-1} and equating input and output terms

$$Y(z) \left[\left(1 + \frac{1}{\omega_a}\right) + z^{-1} \left(1 - \frac{1}{\omega_a}\right)\right] = X(z) [z^{-1} + 1] \qquad (3.48)$$

and applying the shifting theorem gives

$$Y_i = Gx_i + Gx_{i-1} - Hy_{j-1} \qquad (3.49)$$

where

$$G = \frac{1}{1 + \cot (\pi f_c T)}$$

and

$$H = \frac{1 - \cot (\pi f_c T)}{1 + \cot (\pi f_c T)}$$

The filter gain is $A = 1/G$.

Implementation of the difference equation (3.49) is simply a case of suitable sample selection, multiplication by the previously calculated constants G and H, and addition of the product values as described in the previous example.

Although only the design of a low-pass filter using the bi-linear technique is given above, this may be used as a basis for equivalent high-pass, band-pass or band-stop filters by replacing the Laplace function s/ω_c in equation (3.45) by another function of the cut-off frequency, s. A list of these is given in Table (3.1), where ω_c = cut-off angular frequency, ω_l = lower cut-off angular frequency and ω_u = upper cut-off angular frequency.

Table 3.1   **s-plane transformations of low-pass continuous filters**

Required filter	*Replace s/ω_c by:*
low-pass	s/ω_c
high-pass	ω_c/s
band-pass	$\dfrac{s^2 + \omega_l\omega_u}{s(\omega_u - \omega_l)}$
band-stop	$\dfrac{s(\omega_u - \omega_l)}{s^2 + \omega_l\omega_u}$

Further information on this useful method of deriving software digital filter coefficients may be obtained from the references given at the end of this chapter [9, 10, 11].

3.3.3   HARDWARE DIGITAL FILTERS

As we saw earlier, any digital filter implementation consists of a series of multiplications of the input and output sampled signals by constants and the addition of their products. Intermediate products may need to be stored and the logical processes controlled. Thus, the building blocks required for the construction of hardware digital filters are

(i)   memory cells
(ii) adders
(iii) multipliers
(iv) delay units
(v)   control and timing logic.

In many practical applications (i) and (v) will be supplied from external equipment, e.g. a computer or microprocessor. It may not be necessary to use memory cells at all if delay units are available, so that in this case a sequence controller would be required only for (v). The essential units needed to carry out the process of digital filtering are adders, multipliers and delays, and we commence our discussion of hardware digital filters by looking at the ways in which these may be interconnected.

The digital filter can be represented by a linear difference equation as shown in equation (3.27). The rational z-transform of this equation is given as

$$H(z) = \sum_{k=0}^{M} a_k z^{-k} \bigg/ \sum_{k=0}^{P} b_k z^{-k} = \frac{A(z)}{B(z)} \qquad (3.50)$$

Here, we can recognise clearly the three elements of the digital filter, namely: summation of M or P terms, multiplication by a_k or b_k coefficients and delays of k sampling intervals, z^{-k}. The equivalent block diagram is given in Figure 3.12 showing the interconnection of these elements.

However, this is not the only way to synthesise filter operation, and a more economical form is the canonical direct synthesis structure given in Figure 3.13. Here, the common elements are a number of series-connected delay units which are shared by the A(z) and B(z) functions of equation (3.50), each with a separate summer. Mathematically, we note that H(z) can be obtained through the

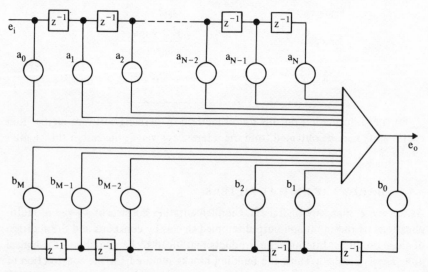

Fig. 3.12   A hardware digital filter

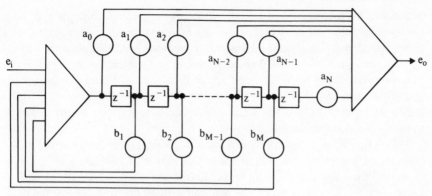

Fig. 3.13   Canonical digital filter

cascade of two filters $1/B(z)$ and $A(z)$, so that denoting w_i as the output of the first filter, the two difference equations for $1/B(z)$ and $A(z)$ are

$$w_i = x_i - \sum_{k=0}^{P} b_k \cdot w(i-k) \tag{3.51}$$

and

$$y_i = \sum_{k=0}^{M} a_k \cdot w(i-k) \tag{3.52}$$

for equal function size $M = P = N$. Hence only one set of N delays are necessary to synthesise this structure. This configuration is for the recursive filter. A similar structure is obtained for the non-recursive filter shown in Figure 3.14. This corresponds to a tapped delay-line or shift register and, although it is simpler, it will require many more delay elements and associated multiplying coefficients to obtain the same filter performance.

3.4   Trend Removal

The acquired data may be accompanied by a trend or baseline modification which will need to be removed at the pre-processing stage. A common trend

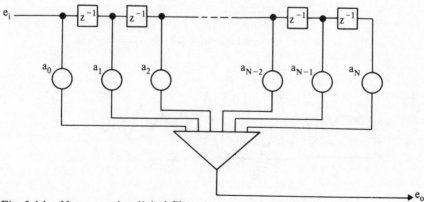

Fig. 3.14   Non-recursive digital filter

introduced by the method of data acquisition or recording is the presence of a constant d.c. shift of the baseline for the recorded variable. This represents a pedestal and may be removed fairly easily, providing that its value is known. If unknown, then the mean may be estimated and subtracted from the signal. It is important to remove this trend before the signal undergoes any integration operation, such as a Fourier transformation, since the error term will be integrated to produce an additive trend which can produce large errors in power spectral density or similar calculations.

Other trends may be introduced by the transducer and are related to its behavioural characteristics. These are in the nature of calibration errors and, providing they are known, can be accounted for in the calibration routine. Large trends of this type should be removed early in the processing of the data, since they can lead to other errors (e.g. amplifier non-linearity due to overloading). A particular induced type of trend generally results where the data is derived from vibration analysis equipment. This has a quasi-periodical nature and may be reduced by the use of tracking filters in the control loop or by subjecting the data to digital filtering. An example of the removal of a low-frequency trend by means of digital high-pass filtering is shown in Figure 3.15. The periodicity of the signal and the nature of its build-up and decay can be seen much more clearly in **(b)** following the removal of the baseline trend present in **(a)**.

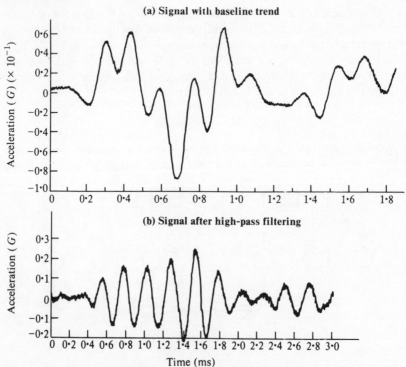

Fig. 3.15   Trend removal by high-pass filtering

It is particularly important in the case of digital data to recognise the presence of frequencies at or close to zero frequency. Analog equipment often does not respond all the way down to zero frequency, so that an apparent filter is present to these very low frequencies. This is not the case with discrete data where frequencies down to zero are inherent in the data. Power at or near zero frequency would be included in the lowest sampling band (one degree of freedom) of width $\delta f = 1/(2T)$, where T is the period of the record. Other sampling bands $2\delta f$, $3\delta f$, up to the Nyquist frequency, $n\delta f = f_N$, will be assumed to contain power samples relative to the recorded signal of interest. Zero frequency (d.c.) shift of the data, or a very low frequency drift, can augment the lowest bands and considerably distort the power spectral density derived from the combined signal. High-pass filtering is used to remove these lowest frequencies, including any slowly varying baseline trend. A cut-off frequency equal to $1/T$ can be chosen for this. A significant trend associated with high-sensitivity line amplifiers is that of additive sinusoidal oscillations at mains frequency and its harmonics. This can occur through inductive pick-up or poorly stabilised power supplies. If it cannot be eliminated at source using screening or other methods, then it may be possible to filter this out from the desired signal using a notch filter.

Other methods of trend removal are curve-fitting to a low-order polynomial in t or a moving average technique. The former is best implemented as part of the analysis digital program and indeed may form a significant processing method. The latter forms more properly a pre-processing operation on the raw data. Here the procedure is to take the deviations of the observed values from a moving average trend. The means of groups of these deviations are taken, adjusted to sum to zero as the estimate of trend. Removal of the trend can then be obtained by subtracting from the original signal x(t). An alternative method has been described in which the ratios of the observed values to trend are taken instead of the differences. An experimental method of removing a background trend is to select the minimum values of the signal on the assumption that these represent the trend alone and to smooth these samples to form a continuous waveform. The inverse of this may then be taken and added to the signal to cancel the estimated trend.

3.5 Decimation

Decimation is a process of data reduction involving the selection of a subset of r samples of the digitised data at intervals spaced uniformly throughout the data sequence. Such data reduction may be necessary because too high a rate of digitisation was originally used or simply because of the limited resolution requirements of the analysis. It is important with digital processing to reduce the quantity of data to be analysed to a realistic minimum, since computing time is expensive and not necessarily linearly related to data length: for example, spectral analysis time could be proportional to the square of the number of points. An example of decimation is given in Figure 3.16, which shows clearly

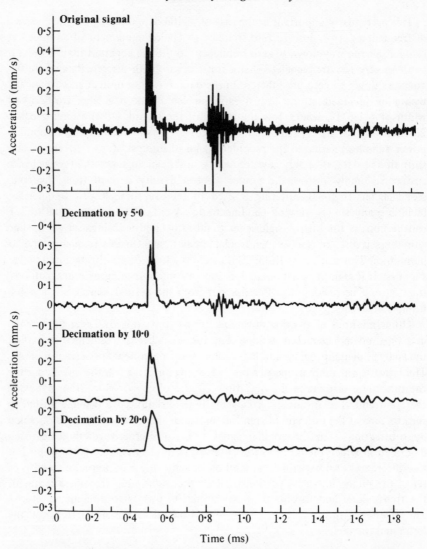

Fig. 3.16 Decimation of a digitised transient signal

the deterioration in resolution obtained with reduction of the total number of points representing the signal. The program necessary to abstract every r'th point from the digital signal is trivial but a decimation program will be more complex than this since it will be necessary to filter the data digitally before decimation. For a sequence of data points, x_i, spaced at uniform intervals, h, this can only represent constituent frequencies up to $1/(2h)$ Hz. If every r'th point is retained, then the new sampling rate will be $h' = rh$ and only frequencies up to $1/2rh$ can be represented. Unless frequencies higher than this are filtered out from the data, then they will be effectively translated (aliased) into the band 0 to $\frac{1}{2rh}$ and thus distort the decimated signal. A recursive low-pass filter to achieve the filtering required was described earlier.

3.6 Calibration Methods

The actual process of recording information from a physical system need not be very complex. The real problem comes in attempting to extract meaningful data from the mass of recorded information that can be acquired. With a properly planned acquisition operation this is a matter of data reduction involving scaling or curve fitting to reduce the raw data to significant values. Where no thought has been given to the relationship between the acquired values and the physical units involved, i.e. a previously considered calibration procedure, then real values can never be ascribed to the data and the best that one can hope for is to obtain relative information between data ensembles obtained at the same time under the same conditions.

Of even greater importance is the need to associate each record with sufficient information in order to identify the record itself, to distinguish it from other records taken at other times in other conditions. It is also necessary to include all relevant characteristics of the measurement itself, e.g. type of transducer, its sensitivity and frequency characteristic, amplifier gain, filter characteristics, sampling rate, etc. Where the data is multiplexed, the order of multiplexing and method of channel identification is required. This is in addition to information on the origin of the data and its measurement environment. We need to know if the data have been affected by local vibration, temperature or humidity conditions, whether visual data have been acquired at a particular angle to the target and many other such details which establish the conditions under which the measurements have been taken.

Calibration considerations begin when the experiment is being planned and, in many cases, calibration hardware may form a major part of the source acquisition equipment, particularly if an automatic calibration procedure is necessary during the course of data acquisition. Calibration will be required to perform the following tasks:

1. Relate the characteristic of the transducer to the parameter being measured, e.g. pressure in pounds per square inch to volts output.
2. Check the linearity, gain, dynamic range, frequency response, etc., of the system following the transducer.
3. Check the changes that may have occurred during the actual period of recording, e.g. amplifier drift, change of transducer characteristic, etc.

We can distinguish three periods when calibration information can be made available:

1. Prior to the recording of the data (pre-experiment calibration).
2. During the recording of data — the calibration information either multiplexed with the signal data or recorded on a parallel channel (real-time calibration).
3. After the data has been recorded (post-experiment calibration).

In addition, we can recognise two forms of calibration:

1. Direct calibration on the transducer within its environment.
2. Substitution methods in which the transducer is replaced by an equivalent source generator.

Pre- and post-experiment calibration is frequently carried out using the substitution method. A typical example is where a transducer is replaced by a highly accurate resistor, through which a measured source current is passed (which can be a.c. or d.c.). The resultant known voltage is recorded, both before and after the desired signal measurement. The absolute value and difference values between the two readings will give information on the fixed gain of the system and its change with time. In a practical case, a series of resistor values would be used to check the dynamic range of the system.

It is generally assumed that the changes occurring during the record are linearly related to the two values taken prior to and after the run. This is not necessarily always the case, and where a non-linear response is suspected then a 'dry run' with the pre-calibration input voltage is also carried out for a duration equal to the record. The continuous record obtained can then be analysed by a least-squares or polynomial method to derive the shape of the system response. The pre- and post-calibration values can then be fitted to the curve and the raw data modified in accordance with the calibration curve so obtained. If the post-calibration procedure is not carried out, it must be assumed that the constants of the acquisition system remain unchanged for the duration of the record. Dynamic calibration of the transducer elements within the framework of the acquisition system is desirable, since transducers are sensitive to environment, and a calibration made in the laboratory under laboratory conditions may differ considerably from a similar calibration repeated under totally different environmental conditions. The general principle is either to operate the transducer as a reverse generator so that it may be driven by an electrical signal and calibrated in terms of its physical change, or to activate the transducer with a known physical change and observe the output voltage or current produced. Our example concerns the calibration of the detection arrangements used for a micro-meteorite investigation carried out in the upper atmosphere using a scientific satellite (Fig. 3.17).

A very thin metallic foil is exposed to the environment through which the micro-meteorite particles are passing. Particles above a certain size are capable of puncturing the foil and leaving a pin-hole, slightly larger than the meteorite itself. Sunlight can then pass through this hole and impinge on a series of photocells, situated quite close to the foil, and which act as transducers converting the narrow beams of light into electrical signals proportional to the area of the pin-hole. An indication is thus given of the size of the meteorite falling on a given segment of the foil. Arrangements are made to move the foil across the detection area at a fairly slow speed, so that continuous information on the quantity and size of the meteorite stream can be obtained by telemetry from the satellite.

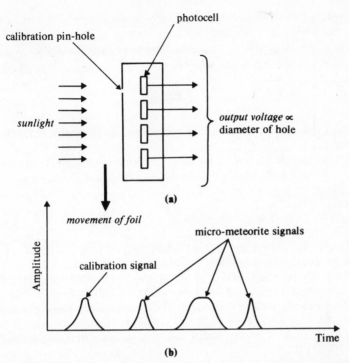

Fig. 3.17   Micro-meteorite detection

Calibration is carried out by including a number of fixed-position holes of a known diameter in certain detection areas. The signals produced by the sunlight passing through these calibration holes can be regarded as a reference for micro-meteorite size, which is independent of photocell ageing or changes in detection and transmission sensitivity, since these are common to both holes made for calibration purposes and those made by micro-meteorites.

Scaling need not form part of the calibration procedure, e.g. where a substitution method is used. It is often carried out later in the main processing program prior to presentation of the final output, in order to simplify the intermediate calculations and so need not form part of the pre-processing operation.

Data identification is important for the reasons stated earlier and, in its simplest form, will consist of a document listing the relevant data which will be retained with recorded information. However, documents are likely to be separated from the relevant data, and it is preferable always to include this information recorded on the same media as the data. Digital coding methods have been developed to permit the identification information, set up on hand keys, or fed in from cards or papertape prior to the recording of the signal information.

Where the data is recorded or converted into digital form then a number of blocks of data preceding the signal data blocks are allocated to contain the identification information, which, in this situation, is referred to as the header information or header label. The data forming this header label would be input from punched cards or punched papertape. In the case of converted data from analog magnetic tape containing its own form of digitised identification, then the digital decoder could be suitably interfaced into the digital computer to enable direct digital transfer to take place.

References

1. CLAYTON, G. B. *Linear Integrated Circuit Applications.* Macmillan, London, 1975.
2. COUGHLIN, R. F. and DRISCOLL, F. F. *Operational Amplifiers and Linear IC's.* Prentice-Hall, Englewood Cliffs (NJ, USA), 1977.
3. ROBERGE, J. K. *Operational Amplifiers, Theory and Practice.* Wiley, New York, 1975.
4. SMITH, J. I. *Modern Operational Amplifier Circuit Design.* Wiley, New York, 1971.
5. TOBEY, G., GRAEME, J. and HUELSWAN, L. *Operational Amplifiers: Design and Applications.* McGraw-Hill, New York, 1971.
6. HENLEIN, W. and HOLMES, H. *Active Filters for Integrated Circuits.* Prentice-Hall, Englewood Cliffs (NJ, USA), 1974.
7. GARRETT, P. H. *Analog Systems for Microprocessors and Minicomputers.* Prentice-Hall, Englewood Cliffs (NJ, USA), 1978.
8. GEFFE, P. R. Towards high stability in active filters, *IEEE Spectrum 7,* 1970, 63–6.
9. CAPPELLINI, V., CONSTANTINIDES, A. G. and EMILLIANI, P. *Digital Filters and their Applications.* Academic Press, London, 1979.
10. CONSTANTINIDES, A. G. Spectral transformations for digital filters. *Proc. IEE,* **117,** 1970, 1585–9.
11. BEAUCHAMP, K. G. and YUEN, C. K. *Digital Methods for Signal Analysis* George Allen & Unwin, London, 1979.

ADDITIONAL REFERENCES

RADER, C. M. and GOLD, B. *Digital Processing of Signals.* McGraw-Hill, New York, 1969.
MILLMAN, J. *Microelectronics – Digital and Analog Circuits and Systems.* McGraw-Hill, New York, 1979.
LUBKIN, J. K. *Filter Systems and Design.* Addison-Wesley, Reading (MA, USA), 1970.

Chapter 4

Data Acquisition

4.1 Introduction

Following translation of the signal source energy into electrical form, further operations will depend on whether the signal is to be processed in real-time, i.e. as it is produced, or whether recording is to be attempted. If the former is the case, it may be sufficient to reproduce the signal in graphical form so that visual inspection can be carried out.

Prior to 1950, most data acquisition systems were analog in form. The classic example of this is the multiple-channel strip-chart recorder used for data logging. This is still in use for much investigatory work, although the advent of the digital voltmeter and low-cost analog-to-digital converters has largely displaced the multi-channel recorder for routine recording purposes. However, for initial evaluation of signal output from a transducer, a visual recording is often required, and providing the frequency of the signal is low enough then a strip-chart recording is usually adequate.

Only in certain cases where the signal contains high frequencies and it is necessary to use a fast-operating reproduction device, such as the cathode-ray-tube (CRT), is reproduction without recording attempted for real-time processing [1]. Recording of the data in some form is nearly always required, either for subsequent analysis, possibly on a translated time scale, or for comparison with recordings made at a different time or under different conditions.

This chapter is concerned with visual reproduction of the information obtained from transducer or other input [2], with the conversion of the data into a form suitable for storage, and with storage of the data. Much of the chapter will be devoted to analog (continuous) methods, since this is usually the original form for the signal [3–6]. For convenience, the techniques of digital magnetic recording will be discussed following a treatment of analog recording, although the subject of analog-to-digital conversion is deferred to a later chapter.

4.2 Ink-on-Paper Recording

We commence our description of recording methods by the slowest of these, namely ink on paper recording. The recording systems are also known as **data plotters** or **chart recorders**. There are two types, the continuous time-recording plotter, also known as the **oscillograph** and the fixed X–Y graph plotter.

The time-recording plotter consists of a roll of paper driven by a motor at a constant speed and having one or more pens writing on the surface of the paper to provide the desired traces. These pens swing from side to side depending on the magnitude of the applied voltage. The trace produced is usually of ink, although a number of other techniques, such as thermal recording, are available.

The X–Y plotter consists of a fixed sheet of paper over which the pen is moved in response to two signal inputs, one moving the pen in the X direction and the other moving the pen in the Y direction. In some, a second pen is included, and occasionally a built-in character generator is found. An automatic changeover mechanism is needed where a second pen is used, in order to switch inputs if the curves start to cross each other.

4.2.1   PEN-RECORDERS

The most widely used type is the **galvanometric** type of recorder. This is the least expensive of the available plotters and suitable for recording a wide range of process variables with an acceptable degree of accuracy. Its principle of operation is shown in Figure 4.1. An electrical signal proportional to the measured value of the variable produces a current in the coil of a galvanometer suspended in the field of a permanent magnet. The moving parts are deflected according to the intensity of current. An elongated arm fitted to the galvanometer carries a pen at its extremity and moves over the chart to record the value of the measured variable. An uncorrected pen would traverse over an arc and in earlier recorders the paper used was overprinted with curved calibration lines. In the modern recorders, the movement is corrected to give a rectilinear trace.

This basic design has been in use for many years despite its obvious shortcomings. Many refinements and modifications have been incorporated to minimise

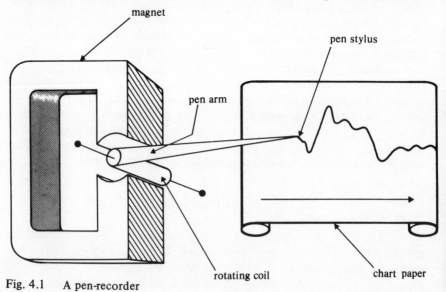

Fig. 4.1   A pen-recorder

these. The most serious is that of pen inertia. The weight of the pen and pen-arm causes inertia and momentum when subject to movement. Both of these are undesirable characteristics, particularly for high-speed recording. Unless special 'penless' techniques are employed then these difficulties limit the speed of recording to less than a few hundred Hertz. Another major problem associated with the flow of ink to the writing tip of the pen plagued earlier designs and caused blotting and smearing to occur, particularly if the pen became stationary for long periods. Now using the modern fibre-tipped pen, a fine trace is produced free from overflow, dripping or clogging of the pen. An alternative design which also produces a clear trace is that of the forced-fluid pen. Here, the pen is pressed against the porous surface of the paper as the pen moves. This gives a clear, dry trace, and the technique is particularly efficient and economical in the use of writing fluid.

Two further modifications of pen recording should be mentioned. One is electric recording in which a low-friction metal wire takes the place of the pen and is caused to traverse over a metalised chart paper. As current flows from the tip of the wire to the chart, metal in the immediate vicinity of the wire is volatilised and a visible trace is produced. The current is interrupted by this process but as the chart moves forward, contact is re-established and the volatilising process continues. In the second method, the pen tip carries a small resistive heating element. Heat-sensitive paper containing a special chemical produces a coloured trace when the heated pen tip travels over the paper. A disadvantage of this method is that the paper is also usually pressure-sensitive so that the chart does not store particularly well.

4.2.2 'PENLESS' CHART RECORDING

Development of galvanometric light-beam recording methods removes the inertia and momentum associated with the pen arm and permits high-speed recording. Typical of these is the **ultra-violet (u.v.) recorder**. This recorder utilises a light source consisting of a high-pressure mercury arc lamp or a quartz-iodine lamp which focuses an intense spot source of ultra-violet energy on to a small mirror carried by a miniature galvanometer. This is then reflected on to photosensitive paper with the light spot deflected across the chart according to the intensity of the electrical signal. The resultant image develops on being exposed to daylight. The very small mass and movement of the galvanometer mirror permit a very high frequency response up to about 80 kHz. Several of these small galvanometers can be mounted side by side to give a number of simultaneous traces which can overlap without the mechanical problems associated with the moving-pen recorder. Improvements in paper speed and control are necessary to accommodate the high speed of recording and, to achieve good control accuracy and reliability, servo-controlled d.c. motors are used. Running costs of the u.v. recorder are high, owing to the necessity of using special sensitive paper.

In another type of penless recorder, the trace is produced by a thin, electrically resistive line, about 250 mm long, formed on a thick film circuit. This is held in contact with the surface of a chart consisting of heat-sensitive paper. Pulses of current can be applied at any of 250 points spaced 1 mm apart along the line. The passage of current raises the temperature of the resistive line at the point at which the current is applied. This heats the surface of the paper very slightly, causing a 1 mm dot impression. Repeated production of these dots in accordance with the magnitude of the signal input forms a continuous trace on the chart.

4.2.3 POTENTIOMETRIC RECORDERS

These are more expensive but also more accurate than galvanometric recorders. The potentiometric recorder utilises a potentiometer or slide-wire to provide pen-position feedback. A d.c. voltage is picked off the slide wire and is compared continuously with the input signal. Any difference activates a servo motor which is used to drive the potentiometer carrying the recording pen arm in such a direction as to minimise the difference potential. This servo action is similar to the feedback seismic transducer described in Section 2.4.4. Precision operation and good frequency response up to about 200 Hz is possible.

4.2.4 THE X–Y PLOTTER

The X–Y plotter consists of a flat bed, upon which a sheet of graph paper is secured, straddled by a gantry which is free to travel in the X direction. A pen carriage is mounted on this gantry along which it is free to move in the Y direction. The pen and gantry are linked separately to wipers of feedback potentiometers, each having a reference voltage applied across them so that the voltage on the wiper will always be proportional to the pen or gantry position. This wiper voltage is compared with the input signal voltage and the difference between them used to drive a servo motor as described earlier. Several independent pens can be carried by the gantry giving a number of independent variables in X which may be plotted against a common independent variable, Y.

It is essential to hold the sheet of paper down firmly on to the flat bed. Whilst mechanical methods have been used, most modern plotters use electrostatic attraction or vacuum systems to retain the chart paper.

4.2.5 CONVERSION OF LINE DRAWINGS

The operation that is the converse of function plotting is sometimes required, namely the derivation of a signal from a graph or line drawing and the conversion of this signal into a digital record for subsequent storage and processing. This process is carried out by devices known as **curve followers**.

A commonly used system is the semi-automatic reader which requires an operator to follow the line drawing with the aid of a hand-held recording pen or crossed-hair device which is tracked by a servo system to provide an electrical output. The reading device, shown as a hand-guided indicator in the Frontispiece, carries a coil energised with an alternating current signal. This is detected

by means of a second coil situated on a moving gantry beneath the plotting surface. The gantry is capable of movement in the X and Y directions in response to
signals from servo amplifiers arranged to drive the controlling motors in such a
direction that the follower coil beneath the table surface always centres itself on
the reading device. In one design of follower, the servo motors also rotate shaft
encoders which provide two sets of binary signals indicating the position in X–Y
co-ordinates referenced to a specified position on the plotting table. After code-
conversion, the digital co-ordinates are available for visual reconstruction or output to the printer, papertape punch or magnetic tape for further processing.

This type of system often incorporates digital display equipment and storage
memory, and operates under the control of a mini or microprocessor from a
standard alphanumeric keyboard. The general-purpose system illustrated in the
Frontispiece is shown diagrammatically in Figure 4.2. The system shown is manufactured by the Ferranti CETEC Graphics Ltd and known as the CDA system.
Details of this are included through the courtesy of this company. A hand-held
cursor digitiser operates in conjunction with a mini-computer supported by disc
backing storage and a visual display screen. This system permits the user to enter
his control and format information through the control keyboard and to monitor
the annotated digitised graphical input on the display screen before finally
deciding to archive the information on to the storage media (disc, paper or
magnetic tape). Many different facilities become available under such a flexible
arrangement. These include plotting of straight lines, arcs, ellipses and rectangles
with reference to datum points taken from the graphical input, parallel lines,
hatching, generation of arrowheads and angled text to accompany the graphical
input. Scaling, rotation and mirror-image transformations are also possible.

Fully automatic equipment requires more complex methods in which the
servo-controlled follower is locked to the recording trace by means of a balanced
signal derived from a pair of photocells situated on the moving gantry. Some
analog methods of curve followers exist where the trace to be followed is pre-

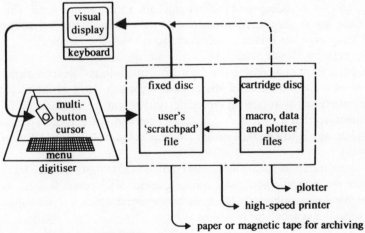

Fig. 4.2   The Ferranti CETEC system

pared using conducting ink which carries the energising signal. This is detected by the gantry assembly situated beneath the plotting surface so that by applying a continuous ramp signal to the X amplifier the gantry is enabled to follow the curve at a linear rate in the X direction and to reproduce a signal at the Y servo amplifier output proportional to the trace waveform. The speed of this type of curve follower is limited to about 1 in/s and is only practicable for short waveforms.

4.3   Analog Instrumentation Tape Recording

For economical storage of continuous signals, the analog instrumentation tape recorder is unsurpassed [3–6]. It is accurate, convenient and flexible in use and its development has achieved a high degree of standardisation between manufacturers. It is still used widely as a storage media despite the fact that, where digital processing of the stored information is required, analog-to-digital conversion has to be carried out upon playback.

An important feature of the analog system lies in the ability to permit linear changes in the duration and frequency of the recorded information during replay in a way not easily obtained with other media. A further advantage concerns the overload characteristics of the analog record. These are less severe with analog recording than other media (e.g. film) so that useful information can be obtained under conditions of poor scaling.

4.3.1   BASIC ELEMENTS OF A RECORDING SYSTEM

For the purpose of description, we can consider a magnetic tape recording system to be composed of three basic elements:

1.   The electronic coding system, which is responsible for the recording process, that is, the encoding of signal information into a suitable form for recording, and for decoding the signal on playback to its original form. This will include signal amplification, automatic gain control, and correction for differing recording speeds as well as the actual encoding electronics.
2.   The magnetic heads, which place the electric information onto tape in magnetic form and convert the recorded signal into an electrical signal of a form identical with that of the original recording. Most of the non-linear characteristics of an analog recording system may be attributable to this element, so that a clear understanding of the conversion process is essential if advantage is to be taken of the inherent low-distortion capability of the system.
3.   The tape transport mechanism, whose function is to move the tape smoothly across the magnetic heads at a constant speed. Malfunction in this area can give rise to a number of distortions of the signal which are peculiar to the process of magnetic tape recording.

Each of these elements will be discussed in the following sections, commencing with the recording process. Many distinct types of coding systems have been evolved for various purposes. We will consider first an amplitude recording technique, known as the **direct recording process**. Although this method is of little value for many signal processing purposes, a study of its characteristics will show clearly the boundary limitations of magnetic tape used as a storage media.

4.3.2 THE DIRECT RECORDING PROCESS

This is a method familiar in the design of domestic tape recorders used for the recording of speech and music. Its chief advantages are the availability of simple techniques of recording and playback, together with an extremely wide recording bandwidth. For the recording of very short transients containing a high frequency spectrum, it will be superior to any of the methods to be discussed later. It does possess, however, a number of disadvantages for signal processing work and, to aid the understanding of these, the method of direct recording will be considered in some detail.

The signal to be recorded in the form of a varying electrical current is passed through the windings of a recording head as shown in Figure 4.3. This consists of a magnetic core in the form of a closed ring, having two narrow non-magnetic gaps inserted at opposite points, as shown in the diagram. The rear gap does not play any significant part in the recording process and its purpose will be described later.

The magnetic path in the head core is completed by the ferrous nature of the coating on the magnetic tape, which is shunted across the front gap formed at the recording head. The signal current flowing through the pair of windings located on each limb of the head core produces a varying magnetic flux across the gap. Due to the non-linearity of the B–H curve for the magnetic material used in the core, the variation in flux will not be proportional to the variation in the recording current so that considerable distortion will take place about the origin, as shown in Figure 4.4.

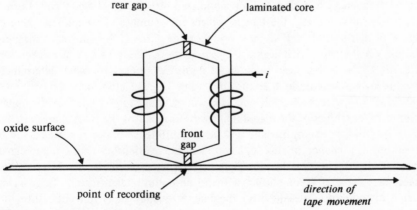

Fig. 4.3   The recording process

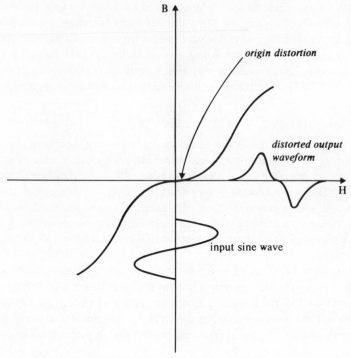

Fig. 4.4   Origin distortion

A simple way of linearising this operation lies in the addition of a d.c. bias current to the signal current, so that only the linear portion of the transfer characteristic is traversed. This method, whilst successful in overcoming the fundamental problem of non-linearity has the effect of reducing the signal-to-noise ratio and has been replaced by an a.c. bias technique which does not suffer from this disadvantage. The method of a.c. bias is shown in Figure 4.5. An alternating current, having a frequency of some three to five times the highest signal frequency, is added to (not modulated by) the signal frequency. During its passage past the gap, the tape is subject to a number of complete hysteresis loops of magnetisation. If there is no recording current, then the mean magnetisation on the tape, as it leaves the gap, is zero. The presence of a recording current displaces this mean value about a positive or negative value, determined by the signal current, so that the final value of magnetisation is linearly proportional to the recording current. This will be understood if we trace the upper and lower envelopes of the biased input waveform across the B–H characteristic. The output flux at any instant will be the difference between curves a and b resulting in a change of flux, c, having a linear relationship to the input signal current. It is important to note that the combination of the a.c. bias and the recording signal is accomplished with no new sum and difference frequencies being introduced. Consequently, the bias frequency does not enter into the recording or subsequent playback processes.

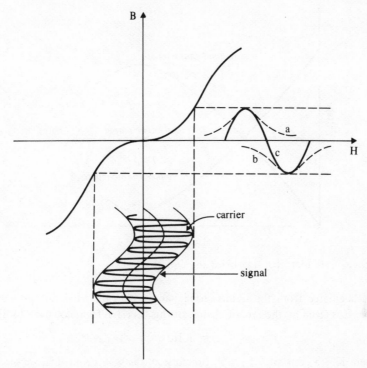

Fig. 4.5   High-frequency a.c. bias recording

The magnitude of magnetic flux established across the recording head gap is proportional to the alternating current through the windings. The actual change in magnetic domains only takes place as the tape leaves the gap area, i.e. on the trailing edge of the gap.

It will be apparent that if the recording current is sinusoidal so that $i = I \sin \omega t$, then the wavelength of the signal to be recorded can be expressed in terms of distance along the tape, permitting a relationship between tape velocity, v, and frequency, f, to be expressed as

$$\lambda = v/f \tag{4.1}$$

The remanence flux will be given by

$$\phi_r = kI \sin (2\pi v/\lambda) \tag{4.2}$$

where k is a constant.

To reproduce the original electrical signal from these variations in remanence flux ϕ_r, the magnetised surface of the tape is drawn past the gap of a reproduce head, similar in construction to the record head. At any given instant, a section of the tape is shunted across the gap and some of the flux surrounding the aligned elemental domains will be conveyed around the head core. The voltage induced in the reproduce head will be proportional only to those flux lines

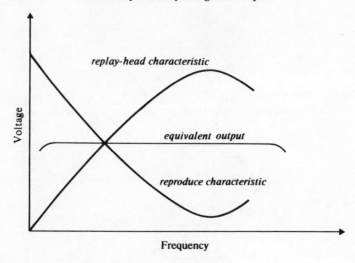

Fig. 4.6   Replay characteristics and equalisation

which emerge from the oxide surface, and not to the total flux present in the tape. It is given by the rate of change of this flux, i.e. from equation (4.2)

$$e = KI\omega \cos(\omega t) \qquad (4.3)$$

where K is a constant of proportionality and is $\ll$ k. From this, we see that the reproduced voltage is proportional to frequency and for constant-current recording the output voltage will have the relationship shown in Figure 4.6. In order to obtain an overall flat frequency response characteristic from the reproduce system, this trend must be removed by incorporating an inverse frequency characteristic, as shown in the diagram. This technique is known as **playback equalisation**. It is also apparent from Figure 4.6 that the signal recoverable from the head decreases with frequency and results in a poor signal-to-noise ratio at very low frequencies where equalisation becomes ineffective. This illustrates a major disadvantage of the direct recording process, preventing storage of signals having very low or zero frequency.

4.3.3   LIMITATIONS OF THE DIRECT RECORDING PROCESS

Some further limitations of this process will be considered below. An important limitation affecting all recording processes is known as the 'gap effect'. As the recorded wavelength approaches the effective gap length then the ratio of the flux linking the core to the flux in the tape decreases. The rate of decrease is described approximately by the periodic function

$$f(x) = \frac{\sin x}{x} \qquad (4.4)$$

where $x = \pi$ gap length/wavelength which reduces to zero when the wavelength

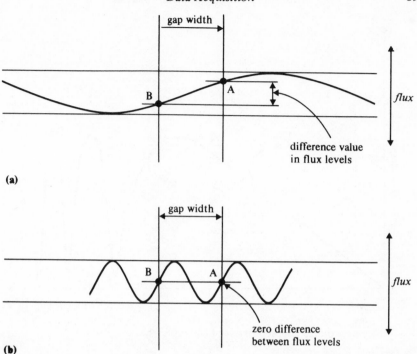

Fig. 4.7 (a) Long wavelength recorded signal
(b) Short wavelength recorded signal equal to gap width

equals the gap length. This effect is shown in Figures 4.7 **a** and **b**. The first figure shows the reproduction of a relatively long wavelength of recorded signal on tape. In the second figure a much shorter recorded wavelength is shown, equal in length to the dimension of the gap itself. Under this condition, the average value of magnetic flux across the gap will be zero and will not change as the tape traverses the gap. The proportional response of the reproduce head, the physical limitations of the gap size, together with other factors discussed below, set bounds to the range of frequencies that can be reproduced by the direct recording process. This is illustrated in Figure 4.8 for one particular tape speed.

A second factor affecting reproduction of high frequencies is the sensitivity of the direct recording process to tape surface irregularities. These cause instantaneous attenuations of the reproduced signal, which are termed **drop-outs**. The effect of these irregularities can be seen from Figure 4.9, which shows the variation in spacing that occurs between the uniform level of the tape oxide coating and the tape head as the tape is drawn past the head. The same effect, of course, can occur due to dust particles adhering to the tape surface. Other limitations of the direct recording process are:

1. Self-demagnetisation: the process of recording effectively causes directional alignment of the magnetic domains contained within the tape material, like

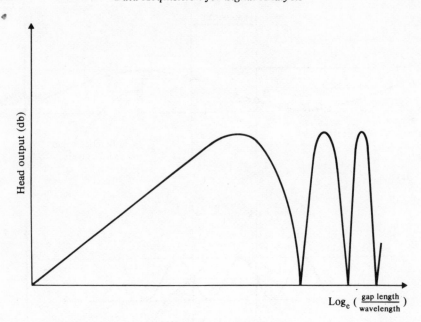

Fig. 4.8   Reproduce characteristics of the replay head

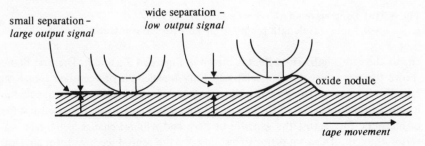

Fig. 4.9   Tape surface imperfections

poles of which are adjacent. For very small wavelengths, the domain strings are short and demagnetisation, due to the proximity of like poles, is considerably increased. This effect decreases as the recorded wavelength becomes longer since the like poles of the effective domain string are further separated.

2. Gap alignment: azimuth tilt can also cause attenuation of the higher frequencies, since for small wavelengths, the induced flux during playback varies along the length of the gap. This has the effect of reducing peak flux value and increasing the minimum value so that the actual flux separation is reduced; the reduction being 100% (i.e. zero induced flux), when the tilt reaches a complete wavelength. This is similar to the effect obtained with a finite gap length and becomes less important as the wavelength increases.

3. Penetration loss: this refers to oxide coat thickness and produces effects which are similar to spacing loss, i.e. shorter excursions of lines of flux from the tape surface occur as the wavelength decreases. With finite tape thickness, as the wavelength is reduced, an increasing proportion of the coating, starting from the inside, fails to make any appreciable contribution to the external field.

4. Recording demagnetisation: the maximum susceptibility of the oxide coating of the tape occurs at a given bias current. The location of this point will, however, vary in depth from the surface of the oxide coating in a manner proportional to frequency, with the result that a reduction in sensitivity occurs as the frequency is increased.

These effects are present with other forms of magnetic recording but, due to the high sensitivity of direct recording to changes in recording wavelength, they are particularly serious with this method of storage.

4.3.4 FREQUENCY MODULATION RECORDING

It has been noted previously that the amplitude performance of magnetic tape systems is subject to a number of limitations, resulting in a degradation of the accuracy for the overall system. In particular, the frequency-sensitivity characteristics of direct recording places a low limit on the minimum frequency that can be satisfactorily recovered from the tape and precludes the recording of d.c. or slowly varying signals completely. Carrier techniques provide a solution to the problem. Using these techniques, the recorded information is contained in the change and rate of change of the carrier frequency. Consequently, saturation recording is often used to obtain the maximum signal-to-noise ratio at the reproduce head, without affecting the accuracy of the information obtained. However, it is not essential to use saturation recording, and there are advantages in recording the carrier frequency using an additive bias level, as with direct recording, particularly for wide-band operation. This is particularly the case where a high packing density is desirable. Using saturation recording, the packing density is limited by the thickness of the oxide coating, whereas no such limitation is applicable to non-saturation recording.

Many techniques of carrier modulation are in use: examples include pulse-code modulation, phase modulation, frequency modulation and, of course, the various forms of digital recording. In practical recording systems, there is little to choose between the use of phase or frequency modulation, and only the latter will be considered, although much of the treatment is also relevant to phase modulation. Various forms of pulse-code modulation have been devised for analog recording and recent developments have indicated considerable advantages over frequency modulation, particularly with regard to packing density. The method will be discussed briefly later in this chapter.

Frequency modulation carrier systems (FM systems) are used extensively in instrumentation and signal processing work to achieve a very high standard of

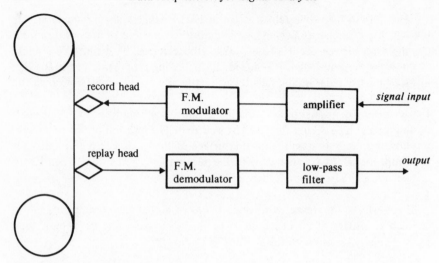

Fig. 4.10   A frequency modulation record–replay system

performance. Here a high carrier frequency is employed which is modulated in frequency by the signal to be recorded (unlike direct recording where an additive carrier is used).

A frequency modulation record–replay system is shown in Figure 4.10. The input signal is applied to a voltage-controlled oscillator to which is connected the recording head. The deviation of the frequency of this oscillator is made proportional to the magnitude of the applied signal. The signal recovered on playback is usually quite small, and must be amplified before being presented to the demodulation circuit. A low-pass filter follows the demodulator in order to remove the carrier frequency and side band frequencies generated during the recording process. This will have a cut-off frequency approximately one-fifth that of the carrier frequency, permitting the recording of signals from 20 kHz down to d.c. at a tape speed of 60 in/s (inches per second). These are typical figures and assume a carrier frequency of 1800 Hz per ins/sec of tape speed which, as we shall see later, is an internationally agreed standard. Recording at a low speed, using a proportionately reduced frequency-range, will increase the recording time available from a given magnetic tape length. If the centre frequency is scaled down in the same way, keeping the maximum percentage frequency deviation constant, then the wavelength recorded in the tape will be the same. This makes it possible to record at one tape speed and reproduce at an entirely different tape speed. The change in signal time base obtained in this way represents one of the most important characteristics of the FM recording process.

Accuracy of performance for a FM recording–playback process is dependent to a large extent on the characteristics of the tape transport mechanism. The coding and decoding electronics will generally have a performance considerably in excess of the minimum requirements for the overall system. The demands on the ability of the tape transport to move the tape across the heads at a precisely uniform speed are indeed quite stringent. Any speed variations introduced into the tape movement at its point of contact with the heads will cause an unwanted modulation of the carrier frequency and result in additional periodic or random components to the wanted signal. It is important to note that the effect of these unwanted speed variations, commonly referred to as 'flutter', will introduce a background noise level which is directly proportional to the deviation obtained. Hence, it is important to reduce these to a minimum, particularly with wide deviation systems, such as sub-carrier FM systems. We can regard these effects as limiting factors which control the dynamic range and accuracy and they will merit further consideration in a later section.

4.3.5 MAGNETIC RECORDING AND REPLAY HEADS

It will now become apparent that many of the limitations of the recording process are intimately related to specific characteristics of the magnetic heads. The construction of a typical magnetic head was shown in Figure 4.3. The two identical core halves are constructed from thin laminations of a material, having both high permeability and low electrical resistance. The latter is required to minimise the effects of eddy-current losses induced in the core, and which increase with frequency. Both halves of the core carry identical windings which are connected in series. Two non-magnetic gaps are shown in the diagram but only the one in contact with the tape enters into the recording process. The purpose of the rear gap is to increase the reluctance of the magnetic circuit such that the heads do not become easily magnetised. This could occur accidentally (for example, due to undesired surges in power supplies), and would result in permanent magnetisation which would reduce the efficiency of the recording and playback process. Since the efficiency of the magnetic head is determined by the ratio of the front gap to the rear gap, the latter is made very small. The gap size of a record head is a compromise between the need to achieve deep flux penetration and high frequency performance. A common value of gap width is 0.005 in. A much smaller gap is used for the replay head.

Multiple-track heads allow several channels of information to be recorded simultaneously. The number of tracks may vary from 2 to 96 and these are accommodated in tapes of varying width, from 0.25 to 2 in. Such a head consists of several cores stacked one above the other. To obtain a good signal-to-noise ratio each track should be made as wide as possible. This conflicts with the need to maintain adequate track spacing and so avoid inter-track coupling or cross-talk. A compromise design inevitably results. Cross-talk can be reduced by using two recording heads and arranging that the even-numbered tracks are recorded or played back by one head and the odd-numbered tracks by the other. Close track spacing on the tape is thus achieved with wider separation between adjacent channels.

Extremely close mechanical tolerances are crucial to obtain optimum performance, using interleaved heads. One of the most vital of these dimensions is that between the gap centre lines of the two interleaved stacks. The two head-stacks are required to be precisely positioned one relative to the other, so that the distance between two lines passing through the centre lines of the gaps is exactly 1.5 in within the tolerance of ± 0.001 in (an international standard). This establishes the relative timing accuracy between information channels recorded on separate stacks. It should be noted that this accuracy is subject to error resulting from tape stretch or shrinkage, which can occur with changes in temperature or humidity. Two other vital mechanical factors are gap scatter and azimuth. Gap scatter denotes alignment of track gap centre lines within the stack. Deviation from the linear alignment is held to less than 100 μin. Gap azimuth refers to the perpendicularity of the gap centre line with a tape surface and is maintained within plus or minus a single minute of arc.

4.3.6 MAGNETIC TAPE TRANSPORT

The tape transport performs the function of moving the magnetic tape at a constant linear velocity across the magnetic heads. It must do this without disturbing the fixed relationship between the position of the recording tracks on the tape and the magnetic head record or replay gaps relative to a fixed datum (e.g. the base plate of the machine).

As mentioned earlier, FM recording is extremely susceptible to variations in transport speed, resulting in a form of signal distortion known as flutter. Two methods of speed control are in current use. The majority of instrumentation systems use a system of capstan speed control in which the output of a tachometer, driven by the capstan, is compared with a signal derived from a stable crystal oscillator. The error signal produced is used to control the speed of the capstan motor so as to maintain this at a precise synchronous speed. To change the capstan speed it is only necessary to alter the frequency division network, interposed between a reference oscillator and the frequency comparison device.

However, this method of control will not necessarily produce a constant tape speed, due to slip between the capstan and the pinch rollers, and also due to the dynamic distortion of the tape. A second method of control is therefore used and referred to as tape speed control. Here a reference signal is derived from the crystal oscillator and recorded on one track of the tape. During playback it is this signal which is compared with the crystal oscillator instead of the tachometer output. With this method of control the capstan speed is permitted to vary, although the actual tape movement is maintained constant to a high degree of accuracy. Both methods of control are used in, for example, Sangamo magnetic tape systems where the replay tape speed is maintained to within 0.001% of the speed at which the tape was originally recorded.

Transport mechanisms are complex and it will not be possible to describe adequately details of their design here. The general principles involved are, firstly to ensure a constant speed of movement for the tape across the record—replay heads, and secondly to avoid any long unsupported length of tape in this area which could give rise to high-frequency fluctuations of tape speed. This is generally achieved by using a closed-loop head configuration, as shown in Figure 4.11. Control of tape tension is obtained by means of one or more biased tension arms supporting the tapes which are free to move, whilst maintaining tape tension substantially constant.

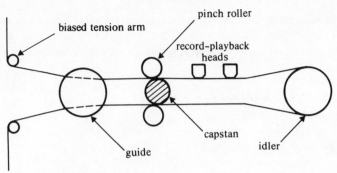

Fig. 4.11   A closed-loop transport system

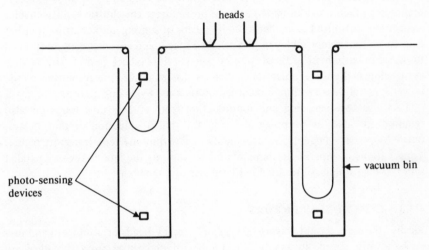

Fig. 4.12   Vacuum-bin system

As an alternative to the tension arms shown in Figure 4.11, there are vacuum-bin storage systems, such as that shown in Figure 4.12. The photo-sensing devices shown in the diagram detect the position of the two loops of tape contained within the bin. They cause signal levels to be supplied to the drive and take-up motors via a servo mechanism, which is arranged to maintain the required tension on the tape, for which the extent of the loop is a measure. This method is used particularly for high-speed systems and is found in digital recording systems, since it also permits very rapid braking of the moving tape.

4.3.7   CONTROL FACILITIES

Operating modes for an instrumentation tape transport can include:

forward movement under capstan control at one of a number of selected speeds
fast forward movement (tape preferably lifted clear from the heads)

forward movement under variable manual control speed (known as the search mode)
Reverse movement under capstan control
fast rewind
manual halt
automatic halt from sensing strip on the tape
loading mode
shuttle operation
loop operation

All of these operational functions are not, of course, found in every tape transport system. Some means of remote control of some or all of these modes is desirable to facilitate automated analysis procedures. The shuttle facility enables a specified section of tape, identified by means of sensing marker strips at either end, to be repeatedly replayed at a selected speed, and automatically re-spooled to the commencement of the section between each playback period. This facility is of value in iterative operations, such as, for example, the resolution of the power spectral density of data recorded on the tape by analog methods.

Facilities for recording and reproduction from endless tape loops are also required for analysis purposes and also for the introduction of time delays. Small loops can generally be accommodated within the tape transport region. Longer loops of tape are accommodated by winding the tape between a parallel series of idler rollers, or preferably by the use of a tape bin.

4.3.8 TAPE MOTION ERRORS

Ideally, the tape should traverse the record–replay heads at a uniform and precisely known speed. In practice, a number of departures from this ideal are found and are attributable to very many sources, not all of which are amenable to design correction.

A shift in the average velocity can be compensated by velocity control of the capstan motor from its servo system, so that precision in tape speed can be achieved fairly easily. Short-term speed variations are another matter, and the control techniques described in Section 4.3.6 can be applied to reduce these. Flutter can take two forms: those in which the flutter variations are identical across the width of the tape, and those which are not. It may or may not be possible to distinguish between the two by measurements of performance. Flutter may be caused by eccentricity in the rotating parts, irregularity due to tension variations, or friction resulting from the rough texture of the tape itself.

Tape traversal speed is thus subject to a combination of effects which can be regarded collectively as a number of individual cyclic variations superimposed on a fairly wide-band random base. The position is still further complicated by differing characteristics of the recording and playback equipment which may not share a common transport. When a signal has been recorded and is later reproduced, the flux circulating in the playback head will be frequency-modulated by the flutter obtained during the recording process, as well as with that present during playback. Since the output from the playback head also produces a voltage proportional to the rate of change of flux then an amplitude modulation component will also be present. It has been shown that the flutter is minimised by choosing a large deviation for a frequency modulation system. Some measure of separation in the recording and playback effects can be made by recording at a high speed (when the effects of flutter are least) and playing back at a low speed. The measured flutter is then almost solely due to that of the replay equipment.

Measurement of flutter for the direct recording process is generally carried out from a r.m.s. sum of the magnitudes of the deviations for all frequencies within a narrow band, up to about 200 Hz, and expressed as a total percentage figure. This approach is not adequate to express the performance of wide-band frequency modulation systems. Instead, a flutter density measurement is made of the percentage r.m.s. flutter obtained over a narrow bandwidth, repeated at contiguous intervals over the frequency range of interest. An alternative method is commonly used, in which a cumulative peak-to-peak curve of the flutter density spectrum is plotted by adding the contribution of successively selected higher frequencies.

Related to flutter is time-displacement error. This is defined as the error between the separation in time of two events, recorded on the same track at the time of recording, and the measured separation time during playback. This particular type of error is cyclic and will attain a number of zero values when the time between recorded events becomes equal to integer multiples of the flutter period. The error will be small for small flutter values and for large time intervals.

A form of timing error occurring between adjacent tracks of the tape is known as skew or yaw error. Here the time or phase displacement takes place across the width of the tape and can have a fixed value or can change with the motion of the tape past the heads. In the former case it may be due to errors in head-stack alignment, incorrect traversal path of the tape across the heads, or phase-shift errors introduced by the reproducing electronics. Variable skew produced by tape motion irregularities is rather more difficult to correct. The actual direction of tape traversal across the head can vary in a non-uniform way and produce a time or phase displacement between tracks which is linearly proportional to the spacing between them. This can be caused by a tension gradient across the tape, which may not be uniform with time. Due to the tensional strains experienced by frequent stop-start operation, the effect is more noticeable with digital transports.

4.3.9 ERROR COMPENSATION

The introduction of a first-order noise component as a result of flutter is one of the major disadvantages of carrier recording systems. However, it is one which can be corrected fairly easily by electrical means. Correction for second-order effects such as time displacement and skew becomes more complex, and both mechanical and electronic methods have been attempted. A simple form of correction for the effects of flutter assumes the generation of coherent errors on all recording channels. Owing to this identity in time, the noise produced by the flutter, in an unmodulated track, can be sign-reversed and used to cancel the effect produced on other modulated tracks. One track of the tape is allocated for flutter correction and an unmodulated carrier signal recorded. No signal should be recoverable from this track during replay, but owing to variations in the speed of the tape during the recording process, the reproduced carrier frequency will vary and produce a reference signal proportional to the flutter variations of the tape. Correction is obtained by sign inversion of this reference and adding it to each of the recorded signals during replay. Under certain conditions, this technique has been successfully applied to earlier types of transport which used a synchronous motor and relied on a heavy flywheel to affect speed stabilisation. The phase differences between the reference signal and the signal being compensated prohibit the simple subtraction of wide-band signals, and, owing to the skew effect, this correction is not so effective for tracks geographically removed from the reference track. The reference track must be located on the same head-stack as the compensated track so that a separate reference track is required for each recording head. The developments in servo control, outlined in Section 4.3.6, have reduced the flutter in tape transport speed to such a low level that electronic compensation methods are now not nearly so effective. This is because the effects of phase differences prevent complete cancellation of the small amounts of flutter remaining, particularly at high tape speeds.

4.3.10   IRIG STANDARDS FOR FM SYSTEMS

In order to allow maximum interchange of information recorded on magnetic tape, using frequency modulating systems, an international standard has been agreed and found fairly universal acceptance. This is known as the IRIG standard (inter-range instrumentation group) [7]. A condensed form of this standard is reproduced below for frequency modulation record—reproduce systems only.

Mechanical characteristics

Magnetic tape: Tape width either 0.5 or 1 in; track geometry as shown in Figure 4.13; track width is defined as 0.050 ± 0.005 in; track spacing is defined as 0.070 in centre-to-centre; track numbering shall be consecutive starting with track one; numbering is taken from top to bottom of the tape when viewing the oxide-coated side of the tape with the earlier portion of the recorded signal to the observer's right.

Head: Two heads are specified for recording and reproduction, with the even-numbered tracks on one head and the odd numbered tracks on the other. The centre lines through the head gaps are to be parallel and spaced 1.500 ± 0.001 in apart. Head-stack tilt is defined by stating that the plane tangent to the front surface of the head-stack at the centre line of the head gaps shall be perpendicular to the mounting plate, to within ± 3 minutes of arc. Gap scatter shall be 0.001 in or less; mean gap azimuth alignment shall be perpendicular to the mounting

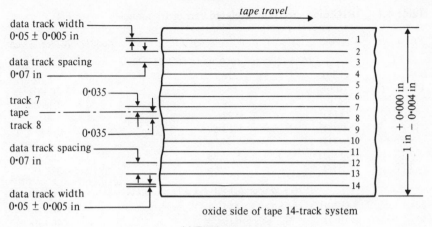

(a) Track geometry

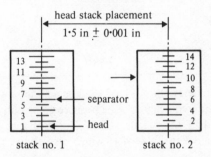

(b) 14-Track analog head and stack configuration

Fig. 4.13   IRIG track and head geometry

plate to within ± 3 minutes of arc; head location shall be positioned within ± 0.002 in of the nominal position required to match the track location shown in the diagram.

Transport: The standard tape speeds are: $1\frac{7}{8}$, $3\frac{3}{4}$, $7\frac{1}{2}$, 15, 15–16, 30, 60 and 120 in/s.

Record–reproduce parameters
Bandwidth: Three bandwidths are designated as follows:

(i)   low band: 0–10 kHz at 60 in/s
(ii)   intermediate band: 0–20 kHz at 60 in/s
(iii) wide band: (group 1) 0–80 kHz at 120 in/s; (group 2) 0–400 kHz at 120 in/s.

Record characteristics: Input voltage of 1.0–10.0 V peak-to-peak shall be adjustable to produce full frequency deviation. Deviation direction shall be such that increasing positive voltage will give increasing frequency.

Table 4.1   **IRIG tape speeds and related carrier frequencies**

Tape speed, cm/s (in/s)		Centre frequency, kHz	Data bandwidth, $\pm \frac{1}{2}$ db
		IRIG intermediate band (± 40% deviation)	
304	(120)	216	d.c. − 40 kHz
152	(60)	108	d.c. − 20 kHz
76	(30)	54	d.c. − 10 kHz
38	(15)	27	d.c. − 5 kHz
19	$(7\frac{1}{2})$	13.5	d.c. − 2.5 kHz
9.5	$(3\frac{3}{4})$	6.75	d.c. − 1.25 kHz
4.75	$(1\frac{7}{8})$	3.38	d.c. − 625 Hz
2.37	$(\frac{15}{16})$	1.68	d.c. − 312 Hz
		IRIG wide-band group 1 (± 40% deviation)	
304	(120)	432	d.c. − 80 kHz
152	(60)	216	d.c. − 40 kHz
76	(30)	108	d.c. − 20 kHz
38	(15)	54	d.c. − 10 kHz
19	$(7\frac{1}{2})$	27	d.c. − 5 kHz
9.5	$(3\frac{3}{4})$	13.5	d.c. − 2.5 kHz
4.75	$(1\frac{7}{8})$	6.75	d.c. − 1.25 kHz
2.37	$(\frac{15}{16})$	3.38	d.c. − 625 Hz

Reproduce parameters: Output level shall be a minimum of 2 V peak-to-peak, with increasing input frequency giving a positive going output voltage. Deviation index should be ± 40% for full deviation. The tape speeds and related carrier frequencies are detailed in Table 4.1, which also shows typical recorded data bandwidths.

4.3.11 TESTING AND CALIBRATION

In the previous section, a brief outline of the IRIG standards was given. How do we set about checking the tape recorder and adjusting the equipment to make sure that these standards are maintained? A number of IRIG recommended test configurations are given in the full specifications; others are given by the instrumentation tape manufacturer. We shall only give a small number of these here. Further information will be found in the references given at the end of this chapter [3–6].

Commercially available test tapes are used for the checking of tape speed. These tapes have saturation-level pulses spaced 1.5 in apart under standard tape tension and other conditions. With the bias and control-track signals disabled, the recorded input is shorted and the tape played. The output pulses are monitored by a counter or timer over a given period and the tape speed computed from this at each speed range of the tape recorder.

Flutter measurements are more complicated. Two test oscillators are required, one for the reference carrier, and one for the two kinds of modulation applied during the testing, namely FM for discriminator calibration and AM to ensure that the system is insensitive to amplitude variations. The principle is to record a precisely known and very stable reference frequency signal on the tape and to feed the reproduction of that signal into a FM discriminator. The output of the discriminator is then calibrated and corresponds to the difference between the record and reproduce speeds.

Three tests that are essential for FM testing are measurement of deviation, centre frequency and polarity. With the input shorted, the use of a frequency counter will check the carrier (centre) frequency of the FM record system. If not within ± 1% of nominal, the frequency should be adjusted to the correct value. To check deviation and polarity, a d.c. voltage equal to the positive full-scale rated input is applied. A check is made that a tape recorded from this signal produces a *positive* d.c. output signal and that the carrier frequency and demodulated outputs are measured to give the correct values. These checks are repeated for a full-scale negative d.c. input voltage. Measurement would normally be carried out using an accurate recording voltmeter. Frequency response is carried out by a second input oscillator with its level set to full deviation (peak-to-full-scale) and recording a number of frequencies spaced throughout the rated signal pass-band to include the maximum modulation frequency. These recorded signals are reproduced and outputs measured from a d.c. or 0 db reference value so that they may be expressed as ± decibel changes. Many other electrical tests will need to be carried out, including bias and record level for direct recording and signal-to-noise ratio, group delay, transient response and linearity for FM recording.

Setting up and testing an instrumentation recorder, particularly a wide-band direct recording system, requires some skill and a fair amount of time. For this reason, automatic methods of testing are preferred and whilst these were originally used only in the laboratory, the application of the microprocessor as a method of control has enabled these tests to be carried out quite easily in the field. This also applies to setting up and running the instrumentation recorder under working conditions. Consider, for example, the controls necessary to adjust the signal electronics in a multi-channel recorder. Controls are required to select the channel on which recording shall take place. The input gain and zero level (for d.c. recording) have to be set for each channel and the reproducing electronics need to be adjusted according to the tape's speed. It must also be possible to monitor the signal, whilst this is being recorded or replayed from the tape.

Figure 4.14 illustrates how a microprocessor can be used to control the recorder transportation subsystem. The servo-control capstan and spool motor drives remain unchanged but are controlled by the microprocessor. This can be provided with coded information to enable it to control the capstan servo mechanism by switching it to forward speed and selecting the frequency from a reference crystal oscillator to set the capstan speed correctly. The status of the deck is relayed to the microprocessor, using a counter to update the tape position. The spool rotation sensors provide information which the microprocessor relates to the rotation of the capstan to compute whether sufficient tape is on either spool. The recording electronics may also be controlled. A manual setting at the control panel may be interpreted by the microprocessor to allow this to

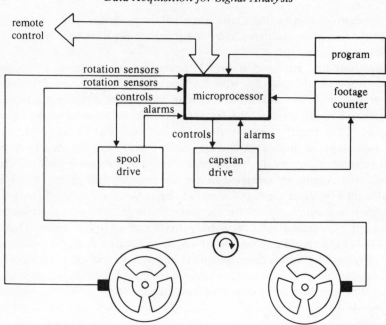

Fig. 4.14   Microprocessor control of tape transport system

provide the correct control signals for each track and to set equalisation to suit the tape speed. In addition, the microprocessor can control calibration signals so that the gain of the recording–replay amplifier may be adjusted, etc. In a system developed by Racal Recorders Limited [8], a number of microprocessors are used to control completely the operation and automatic calibration of the instrumentation recorder and to overcome the problems involved in operating such recorders remotely in the field. This is made effective by a method of connecting the microprocessor, recording–playback electronics, tape transport mechanism and remote panel through the use of a standard well-designed digital interface. This technique, which is not confined to instrumentation tape control only, is described when we discuss complete data acquisition systems in Chapter 6.

4.4   Digital Recording and Storage

In earlier sections, we described the constraints imposed by the use of magnetic tape as a recording medium for analog signals and the methods used to minimise these. At best, the practical limits of absolute accuracy, linearity and stability for analog signals recorded in this way cannot be much better than ± 1.5% over-all input-to-output fidelity when all these sources of error are considered. The problems are non-linearity, noise, drift and dynamic distortion. Digital methods of recording effectively eliminate all of these. This is mainly because digital recording does not involve partially magnetised states. Saturation recording is used, in which the magnetic tape is polarised completely in one direction or the other. However, there is a price to pay, and this is the much lower packing density of information on the tape that is obtained with digital, as opposed to direct or FM recording. Secondly, some simplicity and economy in the record–playback electronics is sacrificed. With digital recording, we need to sample the signal, quantise it and then convert it to a digital code. Finally, the record–playback equipment will require special arrangements to minimise the effect of imperfections on the magnetic tape which become more important with digital recording.

Of course, the initial data may assume a limited number of discrete values and may become available only in digital form. An example of this could be the data derived from solar X-ray observations carried out in the upper atmosphere, in which nuclear counters used for detection themselves quantise the data into small finite levels. We may also have transducers which have a capability of measuring to a degree of accuracy outside the dynamic range of analog storage methods. Digitisation of the signal is then essential and data stored in this form on digital magnetic tapes. However, digital recording is most commonly needed where analog signals are converted, possibly after some analog processing and data reduction, to digital form in order to carry out digital computer operations on the recorded signal. Subsequent storage of the processed data will then be required from data available in digital form. For reasons of speed, convenience, and economy, this storage may need to be carried out using magnetic media, so that a treatment of some of the problems involved in digital magnetic tape recording is relevant to this chapter.

4.4.1 TAPE RECORDING

Digital tape recording is well developed for digital computing equipment and has attained a high degree of standardisation, permitting data exchange between widely scattered installations [9]. The application of this technique is far less critical in performance than analog tape recording and it is not so essential that the user understands the detailed mechanics of its operation. Some characteristics of digital recording are:

1. Considerably reduced dynamic range (only 2 levels of storage required).
2. The data may be read out under synchronous conditions, thus reducing the importance of timing errors.
3. Digital recording is relatively insensitive to tape transport speed variations.
4. The digital process is capable of an extremely high order of accuracy.
5. Record–replay takes place at a high transport speed so that at no time does the magnitude of the signal approach that of the noise level of the process.
6. A lower packing density is employed.
7. Simplified read–write electronics are used, giving high reliability.

Digital information is written on to magnetic tape as a series of discrete areas of magnetisation or **magnetic dots**, along each track of the tape, having a width slightly wider than that of the recording head gap. Each magnetic dot can take one of two saturation states corresponding to a binary 1 or 0. A set of such dots across the tape width is referred to as a stripe, and may be interpreted as a unique coded character consisting of 6 or 8 binary bits. Unlike analog information, the characters are not written or read as a continuous sequence on the tape, but are broken up into groups of characters, referred to as blocks where each block length is separated by an inter-record-gap in which no information is recorded. Block lengths vary but since the inter-record-gap is fixed, the tape utilisation deteriorates as the block length is reduced. This is indicated in the following expression for total real storage capability

$$\text{total number of bits} = \frac{\text{length of tape} \times \text{length of record}}{\left(\dfrac{\text{record length}}{\text{packing density}}\right) + \text{inter-record-gap}} \qquad (4.5)$$

The record length is given in characters, and gap and tape length is given in inches.

The main problems in the use of digital magnetic tape storage in computer-compatible format are caused by the intermittent start—stop operation, consequent upon this block-by-block record—replay process, and the high dependence on tape quality for satisfactory performance. The former results in skew error which is corrected in the design of the recorder by mechanical or electronic means, as discussed earlier.

Digital recording systems are very sensitive to drop-outs, as will be apparent if we consider the method of information storage. Since all the information is contained in the presence or absence of pulses upon playback then the loss of signal or generation of spurious signal by tape inhomogeneities cannot be tolerated. This is a result of the discrete and uncorrelated nature of the recording digital signal. Each signal is accepted as a unique number and not one of a related series, as is the case with frequency modulation analog recording, so that no inherent smoothing is present in the process. We have seen earlier, when discussing direct recording, that drop-outs are most critical at short wave-lengths approaching the size of the replay head gap. This sets a minimum duration for the recorded digital pulse and limits the maximum packing density to a value considerably lower than that obtained with analog tape. Owing to the often catastrophic effects of drop-outs in digital systems and the inability to remove completely magnetic tape surface imperfections, various methods of **parity checking** are used. These usually involve recording an additional digit or digits to indicate whether the sum of the number of 1's recorded are even or odd which can then be checked against a logical addition of the binary bits of the recorded data sample. A check sum of blocks or tracks of data may also be recorded.

The process of digital storage on magnetic tape, although not continuous in the same sense as analog storage, is a continuous process within the blocks of data. It is necessary to provide the data in the form of complete blocks of information at a rate compatible with the speed of transport for a single block. This may, or may not, match the rate of availability of data to be stored, which will, in any case, become available as a continuous character string. Some form of buffer storage designed to hold a complete block is therefore necessary to act as a reservoir to the block-by-block writing sequence. Where the rate of arrival of the data in character form is extremely slow (e.g. when recording of pulses from a slowly decaying radioactive source), then the conventional means of magnetic tape recording becomes inefficient. If the amount of data is small then the information can be recorded on punched papertape for later digital analysis. However, for larger quantities of information, a special form of digital tape recorder, known as an **incremental recorder**, may be used. Here, the tape transport is arranged to move in incremented steps of one stripe at a time. Each input digital character is written as a stripe across the tape, together with a parity bit, and the tape pauses after each character has been written until the next character becomes available and assembled ready to be written. The bit packing density is necessarily smaller than with continuous block-organised tape transports, but the available total storage is considerably greater than would be available from punched papertape, and entry into the digital computer for subsequent analysis is consequently faster. The incremental recorder generally includes the control logic necessary to insert an inter-record-gap at appropriate intervals, so that the resultant tape is compatible with conventional digital forms of storage.

4.4.2 DISC AND DRUM RECORDING

Whilst magnetic tape offers a convenient and economical method of recording and storage for digital information, it does possess the considerable disadvantage of long access-time, particularly where selected pieces of information are required from different parts of the tape.

Disc and drum systems have been developed to improve this situation. Figure 4.15 shows the access arrangements for a magnetic disc storage system. A number

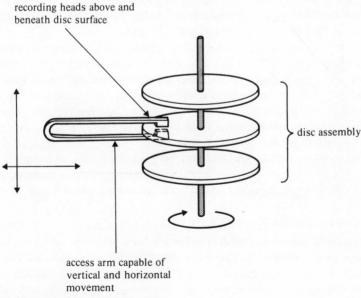

recording heads above and
beneath disc surface

disc assembly

access arm capable of
vertical and horizontal
movement

Fig.   4.15   Magnetic disc assembly

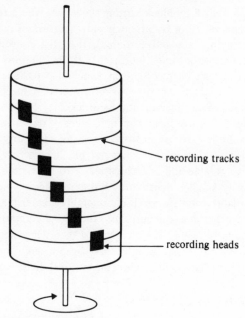

Fig. 4.16   Magnetic drum recording

of rotating discs are mounted on a common spindle, each disc carrying several hundred concentric recording tracks. Access to the stored information is obtained by moving an arm carrying the pick-up heads vertically to the selected disc and then radially to the selected track. A very small head-to-surface spacing (of the order of 0.001 in) is obtained and access to stored information is achieved within a few milliseconds. Storage capacities of several hundred million digital characters are common for digital computer disc pack assemblies.

The magnetic drum also carries a number of adjacent recording tracks over its surface. It differs from the multiple-disc device in that each track has a recording head associated with it (Fig. 4.16). This permits even faster access than with the disc, since it is not necessary to reposition the head each time information is to be read. The storage capacity of the device is less than the disc, since even with the staggered head positioning shown in Figure 4.16 fewer tracks can be accommodated for a comparative recording surface.

Owing to the complexity of the recording head assembly, the drum is considered as a fixed recording media, unlike disc assemblies or magnetic tapes which can easily be changed by removal from the transport mechanism.

4.4.3   DIGITAL RECORDING METHODS

The process of recording digital information must take into account the susceptibility of the method to drop-out errors, mentioned earlier, and the need to contain as much information as possible on a given area of magnetic media (packing density). To some extent these requirements are conflicting and the various methods developed reflect the compromises required.

A brief discussion of the principal methods of digital recording is given below.

Return to zero recording (RZ)
Here the polarity of recording current is determined by each bit of the digital character. For a binary 1 the current flows in one direction and for a 0, the current flows in the reverse direction. Between each digit the current is returned to zero level. This is shown diagrammatically in Figure 4.17. The current flowing in the recording head causes magnetisation of elemental magnets contained along the surface of the tape with a polarity dependent on the direction of the current. Upon playback, the flux changes induced by the passage of the magnetised tape across the reproducing head causes pulses of current to be produced, and these in turn operate to trigger electronic circuits and provide a logic output waveform.

Since a pulse of current is produced for each digit then the signal carries its own timing or clock reference signal.

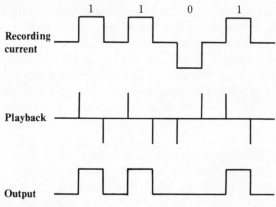

Fig. 4.17   RZ recording

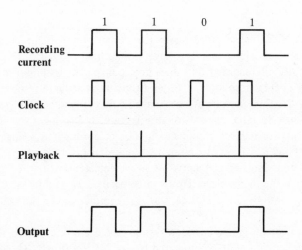

Fig. 4.18   RB recording

Return to bias recording (RB)

A slightly better packing density is achieved with this system where a saturation current level corresponds to a binary 1 and no current to a binary 0. As seen from Figure 4.18, an accompanying clock signal is now necessary, since the precise location of the recorded digits is lost for binary 0's. However, by recording all the bits of a character in parallel on one set of heads, an odd parity signal, necessary for error checking, will also provide a clock signal.

Non return to zero recording (NRZ)

Both the RZ and RB methods are productive of superfluous pulses. RZ requires two flux changes for each digit and RB requires two flux changes for each binary 1. In the NRZ method only one flux change is required for each digit, and better packing density is obtained. This is shown in Figure 4.19. Current flow is reversed only when there is a change in the signal information. Thus a series of 1's maintains a continuous unidirectional flux, and only when the signal goes from one binary state to another is information recorded on the magnetic media in the form of a reversal in direction of the magnetic field. As with the RZ method, a reference clock signal also needs to be recorded.

One important advantage of NRZ over the two previous methods described is that separate erasion of previously recorded information is not required. Since the recording current is always flowing in one direction or another the process of recording automatically overwrites earlier information.

Non return to zero for ones (NRZI)

A difficulty with NRZ recording is that failure to respond on playback to a single recorded digit (change of flux direction) can cause the sequence of ones and zeros to become out of step. In the NRZI system shown in Figure 4.20 every one causes the recording current to change polarity of magnetic flux. Hence a playback pulse corresponds to a binary 1 regardless of detected polarity.

Phase encoding

Both NRZ and NRZI cause problems in the design of a suitable amplifier to reproduce a playback signal that can have a very long sequence of 1's or 0's where the polarity of recorded flux does not change. The amplifier must have a very wide frequency response, in order to include adequate response at low frequencies for the long periods of no change in flux, together with good high-frequency response to rapid changes of flux level. It is also inconvenient to use the NRZ methods for recording on magnetic disc or drum where a single recording track is used and provision of a reference clock signal is difficult.

With phase encoding, each binary bit has associated with it a fixed time period when recording takes place. This is referenced at two points: at the beginning of the time period, i.e. start time, and at the middle of the time period, i.e. centre time.

In one version of the phase encoded system, there are one or two recording current changes depending on whether the binary bit is a 0 or 1. The recording conventions are:

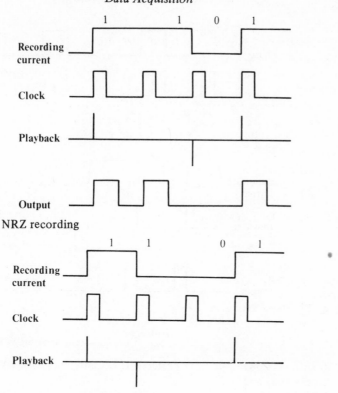

Fig. 4.19   NRZ recording

Fig. 4.20   NRZI recording

1. There is *always* a change of recording current level at *centre time*.
2. The *direction* of change depends on whether a 1 or 0 is present.
3. A change in current at *start time* occurs only when this is necessary to obtain proper change of current at the *centre time*.

This is illustrated in Figure 4.21. The first line shows the binary signal 1101. The corresponding start time current change for 1's is as shown on the second line. However, a current change is always required at centre time which is indicated in the third line. Finally the fourth line shows the actual recording current taking into account the above current change requirements. The playback recovered pulses and reconstituted digital output are shown on the fifth and sixth lines, respectively.

During playback, pulses are produced at every change in recorded flux. These are utilised in conjunction with a clock reference sequence to reconstitute the signal. The clock signal can be generated locally and only requires synchronisation with the centre time when a pulse is *always* available from the signal. For this reason, single-track recording is possible and the system may be used for disc and drum recording.

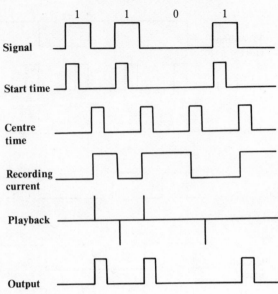

Fig. 4.21   Phase encoding

References

1. *Displays for Man—Machine Systems.* IEE Conference Proc., Lancaster University (England), 1977.
2. MURRAY, R. O. *Computer Handling of Graphical Information.* Conference Proc., Society of Photographic Scientists, Washington DC, 1970.
3. EMI *Modern Instrumentation Tape Recording.* EMI Technology Inc., London, 1978.
4. STEWART, W. E. *Magnetic Recording Techniques.* McGraw-Hill, New York, 1958.
5. DAVIES, G. L. *Magnetic Tape Instrumentation.* McGraw-Hill, New York, 1961.
6. MEE, C. D. *The Physics of Magnetic Recording.* North Holland, Amsterdam, 1964.
7. *Telemetry Standards IRIG Document 106—66.* Secretariat Range Commanders Council, White Sands Missile Range, New Mexico 88002, March 1966.
8. MILE, A and SHORT, M. A. The microprocessor in high-performance recording. *Racalex Lectures,* Racal Ltd, London 1979.
9. FLORES, I. *Peripheral Devices.* Prentice-Hall, Englewood Cliffs (NJ, USA), 1973.

ADDITIONAL REFERENCES

THOMAS, H. E. and CLARKE, C. A. *Handbook of Electrical Instrumentation and Measurement Technology.* Prentice-Hall, Englewood Cliffs (NJ, USA), 1967.
SPITZER, F. and HOWARD, B. *Principles of Modern Instrumentation.* Holt, Rinehart & Winston, New York, 1972.

Chapter 5

Digitisation

5.1 Introduction

The purpose of this chapter is twofold. First, we wish to study some theoretical results concerning the effect of digitisation, namely sampling and quantisation errors. The derivations of these results, included here for the sake of completeness, are rather mathematical, but they may be omitted without loss of continuity. The bulk of this chapter is devoted to the principles upon which the hardware for signal digitisation is based. The special problem of picture digitisation is separately treated in the closing section of the chapter.

5.2 Sampling and Quantisation

As shown in Figure 1.1, digitisation amounts to the conversion of a continuous function of unlimited precision to a set of discrete values of finite precision. The question we need to ask is then, 'How much information has been lost?' or equivalently, 'Under what circumstances is the loss of information acceptable?' By performing such analysis, we aim to digitise a given signal in such a way as to minimise cost without introducing unacceptable errors. That is, we would like to make the interval between successive values, the **sampling interval**, denoted as h, as large as possible so that there is a minimum number of data points per period of time. We also wish to keep n, the word length of the digitised values, short. The problem is thus to relate the error obtained with particular values of h and n.

We shall start by considering sampling alone. Let us ask the question, 'Under what conditions is it possible to reproduce the original function, say $x(t)$, from its sampled values $x(ih)$, $i = 0 \pm 1, \pm 2, \ldots$?' Clearly, if we can reconstruct $x(t)$ from its sampled values, then they must contain all the information originally in $x(t)$, and nothing has been lost. Theory provides a very general answer to this question which, however, is not very useful for practical purposes. It is a result in a more restricted form, the **sampling theorem** of communication engineering, that is useful because the condition it prescribes for error-free function reconstruction may be approximately satisfied for real signals by filtering [1]. The theorem considers a special kind of function reconstruction called **sinusoidal interpolation**, in which we attempt to reproduce $x(t)$ from $x(ih)$ by the use of the formula

$$\hat{x}(t) = \sum_{i=1}^{n} x(ih) \frac{\sin \left[\pi \left(\frac{t}{h} - i \right) \right]}{\pi \left(\frac{t}{h} - i \right)} \tag{5.1}$$

The reason for selecting this formula is that the function $\sin \left[\pi(t/h - i) \right] / \left[\pi(t/h - i) \right]$ expressed in general form as $\sin x/x$, has some very useful properties, so much so that it has a special notation, sinc x. Thus we write equation (5.1) as

$$\hat{x}(t) = \sum_{i=1}^{n} x(ih) \, \text{sinc} \, \pi \left(\frac{t}{h} - i \right) \tag{5.2}$$

An example is shown in Figure 5.1, with $i = 10$. It is seen that sinc $\pi(t/h - 10)$ $= 1$ at $t = 10h$ and is equal to zero at all other sample points, since the coefficient for π will have integer value. Hence the product $x(10h)$ sinc $\pi(t/h - 10)$ will be equal to $x(10h)$ at $t = 10h$ and 0 at all other sample points (see Fig. 5.2). In other words, the product reproduces the function $x(t)$ at one particular sample point without affecting other sampled values. Now consider the sum

$$\hat{x}(t) = \ldots + x(0h) \, \text{sinc} \, \pi \left(\frac{t}{h} - 0 \right) + x(1h) \, \text{sinc} \, \pi \left(\frac{t}{h} - 1 \right)$$

$$+ x(2h) \, \text{sinc} \, \pi \left(\frac{t}{h} - 2 \right) + \ldots \tag{5.3}$$

Here, one product reproduces the value of $x(t)$ at $t = 0h$, the next reproduces $x(t)$ at $t = 1h$, the next at $t = 2h$, etc. Thus an expression similar to (5.2) will reproduce $x(t)$ at $t = 0h$, $\pm 1h$, $\pm 2h$, etc., because each term in it will recover

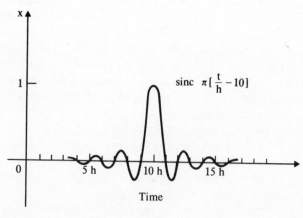

Fig. 5.1   The function sinc $\pi(t/h - 10)$

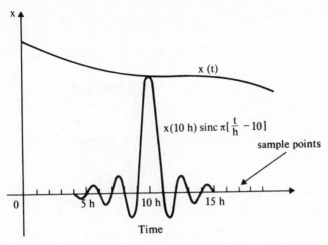

Fig. 5.2 The product x(10h) sinc π(t/h − 10)

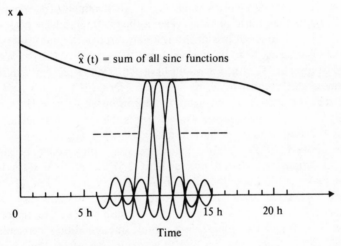

Fig. 5.3 Example of sinusoidal interpolation

one sampled value of x(t) without affecting the other sampled values. The whole process is shown in Figure 5.3.

However, while $\hat{x}(t)$ reproduces x(t) at the sampling points, it does not necessarily provide a complete reconstruction because it may not equal x(t) *between* sampled values. In Figure 5.3 the accuracy seems quite satisfactory, but Figure 5.4 demonstrates quite a different performance. In either case, $\hat{x}(t)$ is called the sinusoidal interpolation of x(t). If we are only given the sampled values of x(t), we can produce some sort of approximation to the values in between (the unknown values), by using the formula (5.2). Whether the approximation is good or not depends on the behaviour of the original function, as illustrated in Figures 5.3 and 5.4.

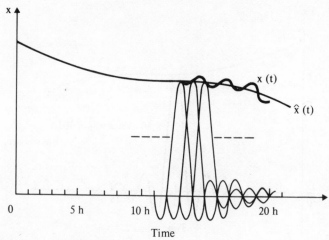

Fig. 5.4   Another example of sinusoidal interpolation

Now the sampling theorem states that if the maximum frequency present in x(t), F, is less than $(2h)^{-1}$, i.e. $F < (2h)^{-1}$, then $\hat{x}(t)$ reproduces x(t) completely. Hence, in order to avoid loss of information through sampling, we must choose a sampling interval, h, to be smaller than $(2F)^{-1}$, i.e. $h < (2F)^{-1}$. With samples spaced at an interval of h, we would have $1/h$ samples per unit time interval. If we denote this number as N, we have $h = 1/N < (2F)^{-1}$, or $N > 2F$. So if we know that the maximum frequency present in our signal is F Hz, we must take at least 2F samples per second. The greater is the range of frequency contained in x(t), the more samples we have to take.

The above is, in fact, only a theoretical result. In practice, because of non-ideal conditions one should take at least 2.5F samples per second. We must realise, however, that even then there is no *complete* recovery of information.

What makes the above theorem useful is that it is fairly easy to limit the maximum frequency present in a signal because we can use analog filters to remove high-frequency components. Thus, the condition demanded by the theorem can in practice be satisfied, at least approximately, There is, however, a limitation, since, according to the theorem, reconstruction of a function using equation (5.2) requires sampled values from the infinite past to the infinite future. Since this is impossible, the condition demanded by the theorem is not fully met, and its application in real life is only approximately valid; however, this in no way detracts from its importance.

5.2.1   PROOF OF THE THEOREM

In this subsection we show brief proofs of the theorem. To follow the discussion, some background in Fourier analysis is required [2, 3]. The proof may, however, be omitted without detriment to the general understanding of this chapter.

We have a frequency limited function x(t). Such a function may be expressed as a Fourier integral of the following form

$$x(t) = \int_{-F}^{F} X(f) \exp (2\pi j ft) \, df \qquad (5.4)$$

We are sampling at an interval of h, with $F < (2h)^{-1}$, so that we may rewrite equation (5.4) as

$$x(t) = \int_{-1/2h}^{1/2h} X(f) \exp (2\pi j ft) \, df \qquad (5.5)$$

because $1/(2h) > F$, while $-1/(2h) < -F$, and $X(f) = 0$ for f outside $\pm F$, so that expression (5.5) includes everything equation (5.4) includes and nothing more. We also note that

$$\int_{-1/2h}^{1/2h} \exp [2\pi j s(t - ih)] \, ds = \frac{\exp \left[\pi \left(\dfrac{t}{h} - i \right) \right] - \exp \left[-\pi \left(\dfrac{t}{h} - i \right) \right]}{2\pi j(t - ih)}$$

$$= \frac{\sin \left[\pi \left(\dfrac{t}{h} - i \right) \right]}{h\pi \left(\dfrac{t}{h} - i \right)} = \frac{1}{h} \operatorname{sinc} \pi \left(\dfrac{t}{h} - i \right) \qquad (5.6)$$

Substituting the above two equations into (5.2) we have

$$\hat{x}(t) = \sum_{i=-\infty}^{\infty} \int_{-1/2h}^{1/2h} X(f) \exp (2\pi j f ih) \, df \int_{-1/2h}^{1/2h} h \exp [2\pi j s(t - ih)] \, ds$$

$$= \int_{-1/2h}^{1/2h} \int X(f) \exp (2\pi j st) h \sum_{i=-\infty}^{\infty} \exp [2\pi j ih(f - s)] \, df ds \qquad (5.7)$$

The summation in the expression has an interesting property. Each term in it is a complex number of magnitude 1 and angle, $2\pi ih(f - s)$. Suppose $f \neq s$, then the sum will always be zero because the terms will be generated in equal and opposite angle pairs so that complete cancellation of the vector-pairs will occur (Fig. 5.5). However, if $f = s$, then all the numbers have the same angle 0, so that they no longer cancel. Instead, they add up to an infinite sum. This quantity may be recognised as a Dirac delta function, $\delta(f - s)$, which is 0 if $f \neq s$ and ∞ if $f = s$. So we have

$$\hat{x}(t) = \int_{-1/2h}^{1/2h} \int X(f) \exp (2\pi j st) \, \delta (f - s) \, df ds$$

$$= \int_{-1/2h}^{1/2h} X(f) \exp (2\pi j ft) \, df \qquad (5.8)$$

Compare this with equation (5.4); we see that they are identical in form. Therefore, when the condition $h < (2F)^{-1}$ is satisfied, the sinusoidal interpolation (5.2) reproduces x(t) exactly.

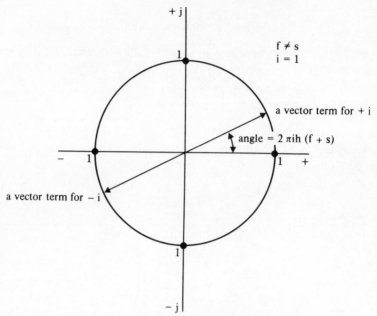

Fig. 5.5   The numbers exp $\{2\pi jih(f - s)\}$ in the complex plane

5.2.2   ALIASING

We said earlier that we can create the approximate condition of making use of the sampling theorem by filtering out high-frequency components in x(t). But what if this has not been done? The answer is that sampling will mix some of these high-frequency and low-frequency components together, making them indistinguishable within the summated waveform. Now if only a very small number of high-frequency components are present, this may not be serious and the result may be almost identical with that obtained by sampling in which suitable filtering is employed.

This effect is called **aliasing**, and to understand this we consider the function $\cos[2\pi(f + N)t]$, with $N = 1/h$. If only sampled values are available we have $\cos[2\pi(f + N)ih] = \cos[2\pi fih + 2\pi i] = \cos(2\pi fih)$.

In other words, if sampling is carried out, the two functions $\cos[2\pi(f + N)t]$ and $\cos(2\pi ft)$ are indistinguishable. Similarly, we have $\cos[2\pi(f + \frac{1}{2}N)ih] = \cos(2\pi fih + \pi i) = \cos(-2\pi fih - \pi i) = \cos[2\pi(\frac{1}{2}N - f)ih]$. Thus, $\cos[2\pi(f + \frac{1}{2}N)t]$ and $\cos[2\pi(\frac{1}{2}N - f)t]$ are also indistinguishable after sampling.

Aliasing is not totally without beneficial effects. Suppose we have a function whose frequency content is limited to between F and F'. Although its maximum frequency is F', it is actually not necessary to sample at intervals of $(2F')^{-1}$, which would be required by application of the sampling theorem. Instead, we may sample at $\frac{1}{2}(F' - F)^{-1}$. This low sampling rate, beneath the limit of 2F', will cause aliasing, by converting some high-frequency components to low-frequency ones. However, since low-frequency components below frequency F were not originally present there is no *mixing* of high and low frequencies, and the information originally contained in the high frequencies can still be recovered, even though it appears to be different. However, we shall not further discuss this interesting possibility because of the need for more mathematics. Readers who wish to know more about this should refer to texts on digital signal processing or communication.

5.2.3 QUANTISATION NOISE

Turning to the effect of quantisation, let us suppose that the input signal has the range of 0 to V volts, and our digital word length is n bits. It is then necessary to divide the voltage range into 2^n intervals and approximate an input within each interval by a different digital value. The intervals may be of unequal size. For example, a Gaussian distribution may be assumed for signal amplitudes. However, more commonly the quantisation intervals are of constant size, as in Figure 1.1; we shall only consider cases such as this. Since the digital output, say x, can only be a number between 0 and $2^n - 1$, the actual value of the analog input is approximately x times $2^{-n}V$. When the intervals are of variable size, it is not so easy to identify the analog value associated with x. One way of doing this is to store the 2^n analog values in a table and look up the element in the position specified by x. This is a fairly complex procedure and is not often carried out. Where the quantisation level intervals are constant then each interval has a size $2^{-n}V$, and any input value whose level is found inside a given interval may be approximated by the mid-point of that interval. Under these conditions the maximum error is $2^{-n-1}V$, and the average magnitude of the error is, approximately, $2^{-n-2}V$. However, from an engineering point of view, it is often desirable to measure the mean square error, which is

$$\int_{-2^{-n-1}V}^{2^{-n-1}V} \delta^2 \, d\delta \bigg/ \int_{-2^{-n-1}V}^{2^{-n-1}V} d\delta = \frac{2}{3}(2^{-n-1}V)^3/(2^{-n}V)$$

$$= \frac{2^{-2n}}{12}V^2 \tag{5.9}$$

Thus the r.m.s. error (equivalent to standard deviation) is about $0.294(2^{-n}V)$. If the signal is evenly distributed from 0 to V, its mean power is $\frac{1}{3}V^2$, so that the signal-to-noise ratio is

$$\frac{1}{3}V^2 \bigg/ \left(\frac{2^{-2n}V^2}{12}\right) = 2^{-2n-2} \tag{5.10}$$

The situation is only slightly changed if the distribution is somewhat uneven.

In addition to sampling and quantisation errors, digitisation equipment also introduces various other errors due to the non-ideal behaviour of the electronic circuits involved. In theory, the design should have taken these factors into account to confine the total errors within very small values. In practice, they can become serious owing to circuit malfunction, degradation through age, or a special combination of components' characteristics which happens to exacerbate the errors. This is why the users should at least be aware of the possibility of such errors. We shall point out typical sources in the discussion to follow.

5.3   Analog-to-Digital Converters

In this section, we provide the reader with a brief account of some of the electronic circuits used for performing digitisation. These are generally known as analog-to-digital (A–D) converters. Numerous kinds of manufactured equipment are readily available to meet almost any conceivable requirement. However, there are a number of technical specifications concerning speed, accuracy and reliability which the users would need to understand in order to ensure system compatibility and satisfactory performance for least cost. We shall discuss common types of digitisation equipment and comment on their important characteristics.

5.3.1   SHAFT ENCODER

Shaft encoders are constructed by making a simple modification to ordinary needle-and-scale meters. They provide a low-cost digitisation method. In place of a needle, a dial with n marked tracks, illustrated in Figure 5.6, is attached to the shaft of the meter. Above the n tracks are placed n detectors, each of which produces a 0 or 1, depending on whether the marking under it is blank or nonblank. Together, they generate an n-bit integer. When a voltage is applied across the meter, the dial turns to a position determined by the value of the input. Now different dial positions place different markings under the detectors, causing a different numerical output to be produced. As can be seen from Figure 5.6, the markings on the dial correspond to binary number 0 to $2^n - 1$, and the further the dial turns, the greater is the integer output.

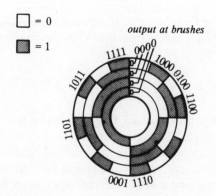

Fig. 5.6   A binary shaft encoder disc

Fig. 5.7   A Gray code shaft encoder disc

The encoder just described has an unfortunate defect; when the movement of the dial places the detectors just over the borderline between two numbers their output becomes very uncertain. For example, when the position is just between 0111 and 1000, each of the digits may be produced as either 0 or 1 because of the uncertainty, so that the output could represent a wide variety of numbers. To solve this problem, markings based directly on binary numbers are not used. Instead, a different system is followed which, however, can be readily converted to binary output. This system is called the **Gray code**. Figure 5.7 shows such a dial. The code table upon which the dial is based is shown in Table 5.1, together with the decimal and binary equivalents. We use four interconnected EXCLU-SIVE–OR elements to convert from Gray code to binary numbers as shown in Figure 5.8. A reference to the truth table given in Figure 1.9a will make its operation clear. The application of Gray code solves the difficulty of borderline values because of the *unit distance* property; successive code words differ in only one digit. When the detectors are on the borderline between 0100 and 1100

Table 5.1   **The Gray code with its binary and decimal equivalents**

Decimal	Binary	Gray
0	0000	0000
1	0001	0001
2	0010	0011
3	0011	0010
4	0100	0110
5	0101	0111
6	0110	0101
7	0111	0100
8	1000	1100
9	1001	1101
10	1010	1111
11	1011	1110
12	1100	1010
13	1101	1011
14	1110	1001
15	1111	1000

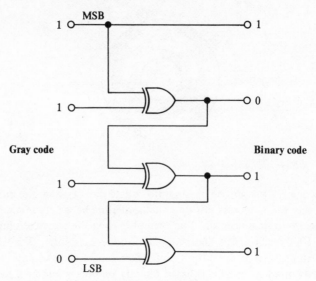

Fig. 5.8   Gray to binary code conversion unit

(corresponding to binary numbers 0111 and 1000), the leading bit is uncertain, and it may be produced as either 0 or 1. However, there is no ambiguity about the other bits. So the output may be either 0100 or 1100, either of which is acceptable (7 or 8, whereas the exact input is 7.5).

Shaft encoders are simple and cheap but their accuracy and speed are both poor. They are now little used, but their principle of operation remains interesting. Also, we shall meet the Gray code again later.

5.3.2   ELECTRONIC CONVERTERS

We have just considered an **electromechanical device**. Now we proceed to purely electronic methods. There are two general classes of techniques: sequential and parallel. In the former, the A–D converter incorporates a **digital-to-analog converter** (D–A converter), and produces an internally generated digital value which is converted to an analog voltage for comparison with the input. This digital value is systematically increased in some way until the equivalent voltage produced from it is almost equal to the input, whereupon the digital value is sent out as the required numerical output. The parallel types of A–D converters employ quite different principles.

So first we must consider the construction of D–A converters. This turns out to be quite simple. Given an n-bit binary number, x, consisting of digits, x_{n-1}, $x_{n-2}, \ldots x_0$, we wish to produce an analog voltage, v_0 in the range 0 to V volts. Figure 5.9 shows the schematic diagram of such a converter. A constant reference voltage, V, is applied through a parallel set of digitally controlled switches, S_{n-1} to S_0, to a set of graded resistors, R_0 to R_{n-1}. These have values $R_i = R_0 \cdot 2^i$, where $i = n-1, n-2, \ldots 0$, so that the currents at the input of the

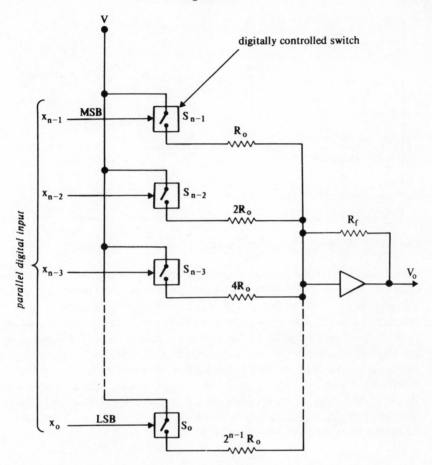

Fig. 5.9   Schematic diagram of a D–A converter

operational amplifier will add in a binary weighted manner and the output voltage will be

$$V_0 = V \cdot R_f \left[\frac{x_{n-1}}{R_{n-1}} + \frac{x_{n-2}}{R_{n-2}} + \ldots \frac{x_0}{R_0} \right] \qquad (5.11)$$

where x_{n-1} to x_0 will have a value 0 or 1 in accordance with the input binary number. Correct scaling is achieved if $R_0 = 2 \cdot R_f$.

The digitally controlled switches are implemented by transistors, and in 'single-chip' fabrication the resistors may be incorporated in the device itself to result in a self-contained unit for D–A conversion.

In several A–D conversion devices, including the sequential ramp-and-hold technique we describe in the next section, we encounter the sample-and-hold amplifier (SHA) as part of the converter. The operation of this amplifier was described in Section 3.1.4. Here we mention briefly the errors caused by the inclusion of the SHA in the conversion procedure. First, we note that the

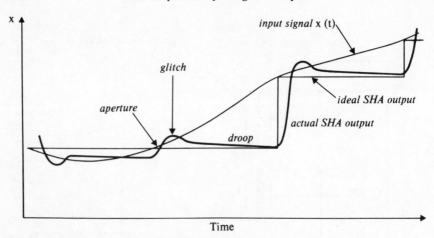

Fig. 5.10   Errors of a sample and hold amplifier

device requires a certain amount of time to stabilise its output during which time the input may vary, so there is some uncertainty about the precise time of the voltage that is actually acquired. This is called the **aperture uncertainty**. Second, the output does not change instantaneously from one sampling value to the next, requiring a certain change time from one voltage level to another. Third, the output tends to exceeed or fall below its correct value momentarily before settling to a steady value. This is often called a **glitch**, which refers to both under and overshoots. Finally, during the period of A–D conversion, there is a slow reduction in output value which becomes more pronounced the less frequent are the sampling periods. This is known as **droop** of the output voltage. These various departures from the ideal output are shown in Figure 5.10.

5.3.3   RAMP-AND-HOLD METHOD

In this method, the analog input voltage in the range 0 to V volts is held constant for the duration of the conversion by means of a SHA unit. An internal voltage (the ramp) is generated. This increases from 0 to V in small incremental steps at a constant rate. The two voltages are compared until the ramp just exceeds the input. The larger is the input, the longer it takes for this event to happen. Consequently, the amount of time taken to reach equality measures the value of the input.

Figure 5.11 shows the actual construction of a ramp-and-hold converter. An n-bit binary counter, which is reset to 0 at the start of the conversion period, increases its count at a constant rate. Its content is sent to a D–A converter. As the counter increases, the output of the D–A converter goes from 0 up to $(1 - 2^{-n})V$. This is the ramp and is compared with the input by an electronic comparator, which produces 0 if the value of the ramp is less than the input but generates a 1 as soon as the ramp exceeds the input value. The logical 1 output triggers the hold mechanism, which permits the content of the counter

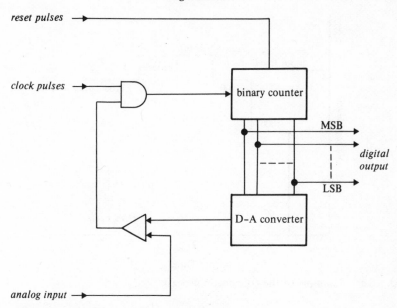

Fig. 5.11   Block diagram of a ramp-and-hold converter

at the time of comparison to be provided as the digital output. Clearly, the larger
the input value, the longer it takes for the digital output to be provided so that
conversion time varies with the value of the input potential. If the input exceeds
$(1 - 2^{-n})V$ equality never occurs, in which case the hold mechanism will simply
cause the final number reached by the counter, namely 1111, to be given as the
output value.

The ramp-and-hold method is, among the electronic methods, quite simple
and of low cost. Further, it does not generate much internal noise because of its
simplicity and regular process of conversion. The main source of error is the
comparator, which often does not respond until the ramp has increased above
the input by some finite margin. Consequently, the digital value may be somewhat too large. However, the real weakness of the method is its low speed; it
takes up to 2^n steps to perform a conversion. Hence it is suitable for only low-
precision (small n) and low-speed conversions.

5.3.4   SUCCESSIVE APPROXIMATION METHOD

This is the most common method for high-performance A–D conversion. It also
employs an internal voltage which is systematically increased until it equals the
input but only n, not 2^n, steps are required. At the commencement of the
digitisation process the value of all the digits comprising x are made equal to 0.
Then we make a rough guess (a first approximation) of the value of the input
by making x_n, the most significant bit of the digital output, have a value of 1.
The number 100 . . . 0 is sent to a D–A converter to produce the internal
voltage, in this case $\frac{1}{2}V$. This is compared with the input. If the former exceeds

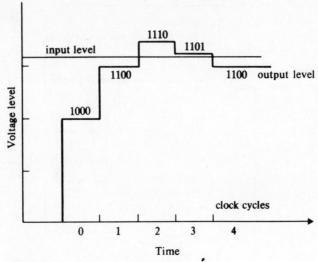

Fig. 5.12   Example of the conversion process

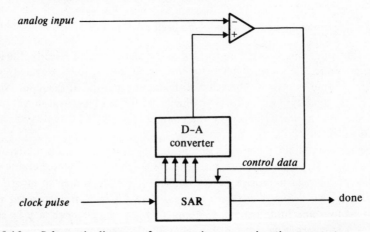

Fig. 5.13   Schematic diagram of a successive approximation converter

the latter, our first guess is too large, and x_n is returned to 0. But if this has not happened, then our guess is not yet large enough, so that x_n remains at 1. (If the two are equal, a possible but rather unlikely occurrence, then the value of x is already correct!) We then make the second digit of x equal to 1. The new guess (either $010 \ldots 0$ or $110 \ldots 0$) is sent to the D–A converter to generate a new internal voltage. Again the 1 is retained if the new voltage generated is smaller than the input, but changes to 0 if it is larger. Then the next lower significant bit is made 1 to generate the next guess. In this way, successive guesses are gradually improved until we have a complete n-bit output digital number accurate to within the least significant digit. Figure 5.12 illustrates what happens during the conversion of a particular value, with the internal voltage rising and falling as conversion proceeds but moving gradually closer to the input value.

Figure 5.13 shows a schematic diagram of such a device. The register marked SAR contains the binary bits approximating to the digital value of the input signal. The contents of this register are connected to a D–A converter, whose output is compared continuously with the input analog value. The results of the comparison controls the unit which inserts or removes a binary 1 from the register. A clock pulse generator provides the triggering pulses for the n bistable circuits required. Initially, the register contains the binary value, 1000. Each clock pulse moves the 1 in the register to the right, so that is located in the bit position to be guessed during the current step, but may be reset to 0 if the comparator output is high.

Figure 5.14 shows the internal construction of the component at the bottom of Figure 5.13, the **successive approximation register (SAR)**. This incorporates a particular type of bistable or flip-flop circuit which has an additional input to which a repetitive clock pulse signal is applied. This is known as a JK flip-flop, the symbol for which is shown in Figure 5.15a and its truth table in Figure 5.15b. The JK flip-flop differs from the set-reset flip-flop in the *time* of change for the bistable which is determined not by a change in the J and K inputs by themselves but by the regularly spaced pulses applied to the clock input terminal, C. The change of state for the bistable occurs coincidentally with one of these clock pulses, assuming that the necessary level has been applied previously to the J or K input terminals. Unlike the set-reset flip-flop, equality of values for the two inputs is permissible but will not result in a change in the bistable state. This clocked type of flip-flop is essential for many computer or counting circuits where operations are controlled on a regular step-by-step basis through the use of a clock pulse generator. In the case of the SAR of Figure 5.14, the clock pulses are applied to all of the flip-flops at the same time.

Initially, the register is cleared by the START(STT) pulse. The first clock pulse would shift a 1 into the most significant bit position (MSB). The value in the register, 1000, is converted into analog form by the D–A converter, and compared with the input. If it already exceeds the latter, the control data line carries a binary 1 which causes the 1 in MSB to be suppressed. Otherwise it is allowed to remain. The second clock pulse then shifts a 1 into the next bit position and the new value in the register, either 0100 or 1100, is compared with the input. This continues until the value of all the bits has been determined. After the fifth clock pulse, the output from the final bistable, shown as DONE, is turned to a 1 to indicate that the conversion process is complete, and the digital value may be read out. A more detailed explanation for the operation of the successive approximation register has been given by Anderson [4]. Alternative designs, using OR gates rather than AND gates, are possible [4, 5].

As in the ramp-and-hold method, the comparator tends to be the critical element. However, because of the more complex operation and the necessity to increase and decrease the internal voltage several times, the device generates more electric noise and glitches than the previous method. The successive approximation method is fairly fast and has the potential for high precision.

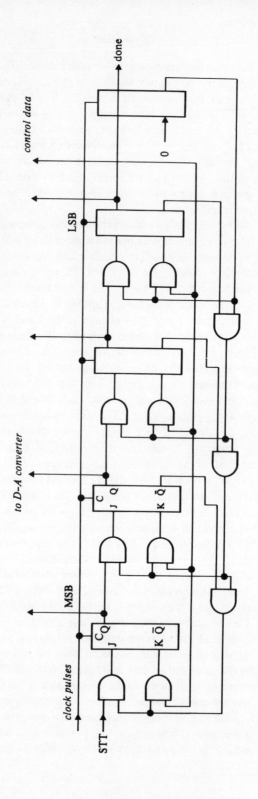

Fig. 5.14 Successive approximation register

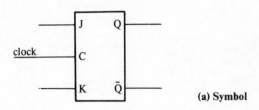

(a) Symbol

J	K	Qn + 1	Q̄n + 1
1	0	1	0
0	1	0	1
0	0	*no change*	
1	1	Qn	Q̄n

(b) Truth table

Fig. 5.15   The JK flip-flop: Qn + 1 is the output immediately after clock pulse; Qn is the output immediately before clock pulse

5.3.5 PARALLEL METHODS

The techniques described in this subsection are capable of even greater speed than the successive approximation method, but for various reasons the output word-length tends to be limited. Whereas the previous technique requires n sequential steps controlled by a clock, here the operation of each method is, in some sense or other, parallel. Also, Gray code is prominently featured here, whereas both of the previous two methods produce direct binary output.

Figure 5.16 shows a 3-bit, fully parallel, analog-to-Gray code converter. On the left is a network of resisters which accepts the reference voltage V and produces a series of seven internal voltages $i/8(V)$, $i = 7, 6, \ldots, 1$. The input is compared to every one of these. If it is smaller than the lowest internal voltage or larger than the highest, then the output should be 000 or 111, respectively. Otherwise, it must fall somewhere between two interval voltages, say i and $i + 1$, in which case the comparators 1 to i would output 1, and $i + 1$ to $i + 7$ would generate 0. A network of OR elements then converts the comparator outputs to the Gray code representation of i. The reasons for this particular set of interconnections will be left as an exercise in Boolean algebra for the reader (see Section 1.7.1).

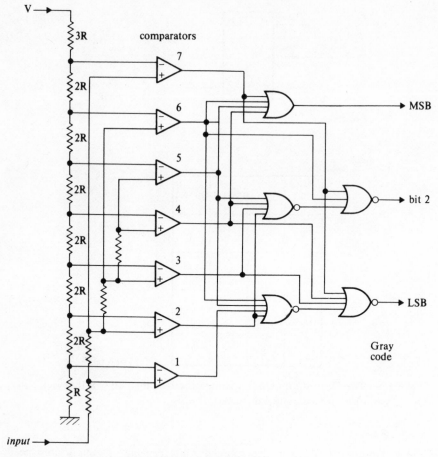

Fig. 5.16   Parallel analog to Gray code converter

The device is extremely fast, but requires $2^n - 1$ internal voltages and comparators, so that the system becomes very elaborate indeed for n greater than 3. Figure 5.17 shows a different method, which produces Gray code output directly from the analog input without first deriving the equivalent binary code. It may be shown that, apart from the first stage which simply establishes whether the most significant digit is 0 or 1, subsequent digits are obtained by subtracting the output provided by the full-wave-rectifier (fwr) (absolute value of its input quantity) from a fixed binary level. An explanation of this method and the theory behind it is given in another work, to which the reader is referred for fuller details [6]. The device is somewhat slower and cheaper than the previous design, since it requires n, rather than 2^n components, to perform subtraction and rectification, and information has to flow through n stages. It does have the problem that errors in successive stages accumulate, so that n must be kept rather small. A value of 6 is a reasonable practical limit.

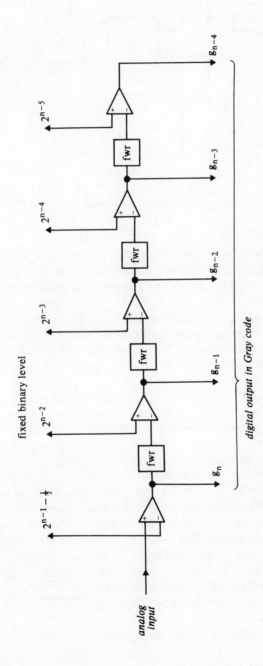

Fig. 5.17   Cascaded analog to Gray code converter

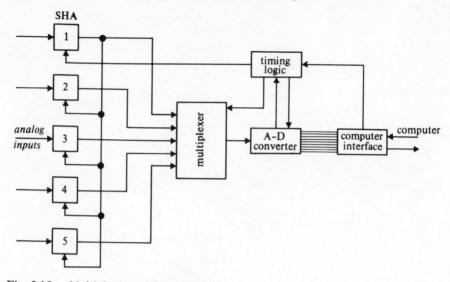

Fig. 5.18　　Multiplexing with simultaneous samples

5.3.6　MULTIPLEXING

In many applications, multi-channel data acquisition is carried out either from multiple data sources, or from the same data source which, however, contains a number of variables that are of interest. It is often impracticable to handle each individual variable with a separate A–D converter because this would increase the amount of interface and I–O control work in the processor. In consequence, it is common to **multiplex** the analog channels, i.e. to make them share the same A–D device. This is especially useful when the analog variables are of the same type, though this is not essential in order to take advantage of multiplexing.

A multiplexer is simply a switching unit having a number of input channels and a single output channel, which is connected in turn to each of the individual input channels. One such configuration is shown in Figure 5.18. Here several low-rate sample-and-hold units are multiplexed to share a common, high-speed, A–D converter. Each SHA will maintain its output constant for duration NT where N is the number of input channels and T is the conversion time of the A–D unit. During this period, the A–D converter successively performs N conversions, one for each input channel, and delivers the digital values to the computer, which will later separate them and sort the values back into N separate groups. The timing logic is set by the processor through the interface and carries out overall control in the process. At the start of each period the logic control sends a common pulse to all SHA units which causes all the analog inputs to be sampled. It then sends N successive pulses to the multiplexer which connects the SHA outputs in sequence to the A–D unit and initiates the N conversions.

Figure 5.19 shows a cheaper arrangement. Here there is only one SHA unit.

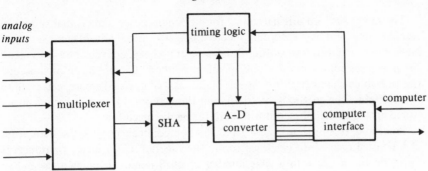

Fig. 5.19   Multiplexing with a shared SHA unit

The multiplexer connects the N analog variables in turn to the SHA device, whose output is converted. Two things should be noted. First, the SHA unit must now work at a higher rate, taking a sample for every duration T, not NT. Second, the different channels are sampled at different moments and not all at the start of the period, as shown in Figure 5.18. Thus the information acquired from the channels is not simultaneous, which may be unsuitable for some applications.

Yet another possibility is to have individual A–D converters for the channels and to multiplex the digital outputs using a digital multiplexer. This used to be considered a high-cost alternative, although it does have the advantage of high rate of working and simultaneous sampling of all the channels. However, with the recent development of 'single-chip' and hence low-cost A–D converters having adequate accuracy, this is now becoming the preferred solution for many applications, not least where data are measured at disparate locations and a minimisation of communication errors is desired. (See Chapter 8 for a discussion of such systems.)

A cyclically switched multiplexer assumes that all the input channels operate at the same data rate. Where this is not true, more complex switching processes will need to be adopted. In particular, one can have a software controlled multiplexer which is switched to the channel chosen by the program upon the output of the channel number. Such systems are, however, rather complicated and elaborate and will not be discussed further here.

5.4   Picture Digitisation

Conversion of pictorial information into a stored digital form requires specialised equipment which is often highly dependent on the original form of the visual data. Before we consider the actual process of digitisation it is therefore necessary to consider some of this equipment and the methods used to obtain a sampled and quantised version of the picture [7]. The picture requires quantisation not only in terms of grey level, which is equivalent to the magnitude quantisation of a single-dimensional variable, but also in terms of co-ordinate information and often of colour as well.

The most common original source for picture or 'image' data is that of photographic film. Film as a recording medium possesses many advantages over other methods of acquiring and storing visual data. It is cheap, convenient and in many circumstances represents the only possible method. Examples are oscilloscope traces, bubble-chamber events, X-ray plates, as well as visual scenes, e.g. satellite photographs. The methods of translating film transparency information into a stored analog or digital record will therefore be discussed first.

Processing of film transparency information commences with the translation of the visual image into electrical form. The function of an image transducer is to divide the image into a large number of small non-overlapping areas and to convert each of these picture elements, known as **pixels**, into a quantised value corresponding to the optical density at that point. The basic characteristics of the image transducer are:

(i)　spatial resolution: the number of individual elements which can be individually examined

(ii)　intensity resolution: the number of grey levels which can be determined

(iii)　speed: the time required to select and measure a point.

Since the processing is invariably carried out using the digital computer, these transducers, which produce essentially an analog output signal, are usually associated with digitising equipment and means of retaining the digitised image information on a storage media such as magnetic tape. Digitisation is considered later in this chapter. Here we will be concerned with the electrical analog of the image for input into the digitiser.

The image transducer consists of three primary elements: a light source, the transparency or sample object, and a light measurement device. A means is also required for selection of a single picture element for measurement. Any one of these three primary elements can be moved to cover the complete picture area for conversion.

For convenience we can regard the complete picture or image to be subdivided into a large number of sub-areas or pixels. Each pixel is represented by one sampled area in which the average intensity level over the area is transmitted or recorded. We consider first a moving light source which scans the complete picture area and allows the transmission in sequential form of a signal representing the point-by-point density of the transparency.

5.4.1　MOVING LIGHT SOURCE

The most common device utilises a moving light source of sufficiently small size to transmit sequentially information about each pixel of the image. A widely used electronic transducer of this type is the flying-spot scanner. This uses a moving spot of light derived from a cathode-ray tube to scan the picture area.

Early flying-spot scanners were capable of recording only the outline of the subject, made available on transparent film. Only two gradations in brightness

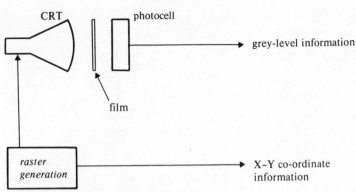

Fig. 5.20   A simple flying-spot scanner

could be registered so that the method was suitable only for line drawing, out-lines and photographs taken from CRT traces. A simple flying-spot scanner is shown in Figure 5.20. A spot of light generated in a cathode-ray tube is caused to traverse a television raster pattern over an area corresponding to a film trans-parency held close to the CRT face. A photo-detecting device collects the light passing through the film which has been modified by the light-transmission properties of the image transparency. Tube deflection errors may be removed by referencing the deflection to a fixed-position graticule. In one design, the flying-spot scans the film transparency and graticule simultaneously. The graticule pulses are counted and the number of pulses reached at the image edge position gives a measure of the Y co-ordinate. The X co-ordinate is obtained in a similar way with reference to a second graticule. The co-ordinate information, together with the image density in terms of full or partial transmission of light, are thus obtained and available for storage and subsequent processing.

Typically, modern flying-spot scanners or film readers are able to record 64 grey levels information which are recorded, together with the co-ordinate infor-mation concerning the image beng scanned. A typical design is shown in Figure 5.21. To measure precisely the relative density of the transparent film at any point it is essential to compare the amount of light actually displayed on the CRT with the amount passing through the film. This is accomplished through the use of the beam-splitting mirror, shown in Figure 5.21, which directs the light into two paths. One path goes through the film and impinges on the photo-detector unit, A. The second path is deflected directly on to a second photo-detector unit, B, via a lens system. The outputs of the detectors are compared and the difference gives a measure of the grey scale intensity. This method, of course, gives only relative values of light transmission. Absolute measurements are made using a different system, known as a microdensitometer, which is considered later.

Two modern developments in moving light source scanners deserve mention. The first utilises a laser light source and is capable of very high resolution, permit-ting scanning speeds in excess of those achievable using a CRT. Scanning is achieved by positioning the very fine laser beam on the film with a light-weight

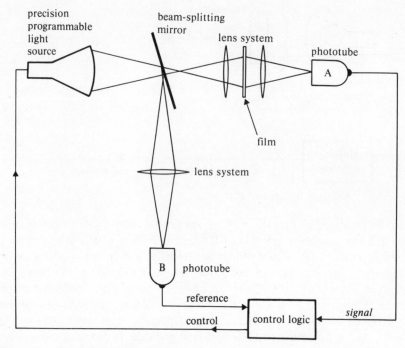

Fig. 5.21   A complete film reader

rapidly rotating mirror. The disadvantages of the method are cost and loss of flexibility, since, as with most electro-mechanical devices, a fixed scanning pattern is required.

A second recent design utilises a solid-state light source in the form of a grid of tiny light image elements. These cover the complete area of the image transparency and permit up to 256 levels of light intensity to be recorded. The location of the light source is constrained to follow a raster pattern, as with the flying-spot scanner, and is controlled by a computer which also registers the X–Y location of the light source in use of the recording media. The resolution and precision of this technique are high, allowing writing speeds of 100 mm/s to be realised with writing spot sizes of only a few microns.

5.4.2   USE OF TV CAMERAS

A less precise but cheaper alternative to the flying-spot scanner is to use a television camera tube as the optoelectronic transducer. A number of camera designs are available which differ in features offered, such as resolution, sensitivity, etc. A popular camera is the Vidicon where the light from each picture element causes an electronic charge to build up at the corresponding point on the camera image screen. The signal output depends on both the incident light and a charge accumulation time. Hence, to remove this dependence on time, the Vidicon tube must be scanned at a constant rate. The target upon which the image falls is a photo-conductive mosaic which is deposited on a transparent metal film. Light passes through the signal plate and falls on this mosaic changing its resistivity. An electron scanning beam traverses the mosaic causing a current to flow which is related to resistivity value and hence the amount of light falling on the small area under the scanning beam.

An improved version of the Vidicon is the silicon Vidicon. This is similar in operation but has a different design of target. This is a silicon wafer consisting of a closely packed matrix of diodes facing the electron scanning beam. The diodes are back-biased so that the light photons falling on a particular diode produce electron–hole pairs. Scanning the wafer with the electron beam will release a current proportional to the number of photons falling upon it and hence brightness of the scene. A high current output is produced and a wider spectral response is obtained compared with the earlier Vidicon designs. A slight modification in design of the silicon Vidicon is known as the Plumbicon. The name refers to the type of diode used which is constructed from lead monoxide. Compact design and high sensitivity are a feature of these Vidicons.

Most television camera tubes provide an output current approximately linearly dependent upon the intensity of light falling upon it. For example, the silicon Vidicon tube has the characteristic expressed by

$$I = kW^\gamma \tag{5.12}$$

where k and γ are constants and W is the light intensity. The gamma constant, γ, is important and, in this case, is approximately equal to one. This may not be what is required to match the subjective performance of the human eye which has a response characteristic logarithmic in form. If we use a logarithmic scale instead of a linear one, this also allows the computer to distinguish a much larger range of intensities than the uncorrected linear response. Most television cameras, therefore, have 'gamma correction' which alters the effective gamma value to about 0.4. The noise characteristic of gamma tube systems can be inferior to other image conversion devices, as the curve $W^{0.4}$ approximates the function $\log_e W$ quite well. Three major sources of noise are:

(i)   shot noise in the imaging device
(ii) thermal noise arising from the load resister used
(iii) amplifier noise.

Some pre-processing may be required to reduce these noise elements to an acceptable level.

Solid-state equivalents to the Vidicon are the various charge-coupled devices making use of MOS techniques. These do not require an electron scanning beam and associated equipment so that the camera can be made very much more compact. A matrix of semi-conductor elements forms the target upon which the scene to be recorded is focused. In each of these elements a charge is created by photon bombardment and stored in a potential well formed by each element in the semi-conductor matrix. This is known as the primary well. The operation of the tube causes transfer of these charges into a secondary storage well and eventually out to a digitiser. The action is very much like that of a shift register found in computing equipment in which a separate shift register is regarded for each pixel. As the image falling on the semi-conductor matrix is sampled, so the charged is moved one stage along to the secondary well and to the digitiser, allowing the value stored in the primary well to acquire a new stored value related to the new picture sample.

5.4.3 MOVING SPECIMEN

Devices which move the image specimen between the light source and detector are known as microdensitometers. They are exclusively mechanical and are amongst the earliest of image transducer designs. One type functions by stepping a glass slide in a raster pattern whilst the light passing through a small aperture to a photomultiplier is measured. Rotating drum scanners can also be used. The technique is only of value with certain types of image, since microdensitometers are slow and inflexible in use. They possess advantages in terms of accuracy, however, since the well-calibrated light source and photodetector used permit the calculation of the percentage of the incident light transmitted through the film plane on an absolute basis. Resolution can be as little as 1μm in spot size.

5.4.4 DIGITAL CONVERSION

The procedures for digital conversion of a two-dimensional image are similar to those described in the preceding sections concerning the digitisation of single-dimensional information. We need to sample the picture area using a procedure compatible with retention of sufficient information for picture reconstruction. We need to quantise the amount of picture brightness into a finite number of discrete grey levels and finally this sampled and quantised data will need to be coded in a recognised digital pattern. In more sophisticated environment, it may also be necessary to take into account colour differences and depth of picture, i.e. three dimensional form [8, 9].

We have considered the various methods of scanning the image to give an electrical output of the form required. There are two major difficulties encountered with two-dimensional digitisation compared with its single-dimensional counterpart: first, the quantity of data acquired from a single digitised picture could exceed 10^7 bits, for even a modest degree of reconstructional accuracy. Secondly, where the picture, reconstructed from this digital information, is viewed subjectively by a human scanner, then the noise content becomes of prime importance. 'Noise' is meant here as any extraneous and irregular information included in the digital representation but not present in the original picture. It thus corresponds in effect to its audio equivalent from which the term is borrowed because of its widely understood and apposite interpretation. As a very general observation, it has been noted that whereas a signal-to-noise ratio as low as 15 db can be tolerated in relation to a sound signal (e.g. a telephone conversation) the corresponding figure for picture information may have to be as high as 40 db (equivalent to an amplitude ratio of 100 to 1).

We consider first the general methods used for sampling and quantisation following the scanning procedures described earlier and complete this chapter with some consideration of the noise problem.

5.4.5 SAMPLING AND QUANTISATION

The basic method of picture sampling is to select a finite sample at pre-selected points over the surface area of the picture and to define a reconstructed picture by interpolation between the grey levels found at the selected points. This procedure will be modified by the mechanism used to carry out the sampling process as noted earlier and it may be more convenient to divide the picture into a number of pixels and to take the average grey level of the pixel as the sampled value. Since interpolation in two variables is a complex process it is usual to choose sample points or areas along a set of 'scan lines' similar to a television raster and to interpolate using functions of a single variable along each line.

As stated earlier, the process of quantisation transforms the magnitude (brightness) of the samples image into a discrete number of steps (grey levels). Each grey level in the sampled picture is replaced by the discrete level closest to it. 32 grey levels corresponding to 5 digital bits of information are usually adequate for most processing purposes. Even spacing of these grey levels is not necessarily the ideal situation in the case of picture processing. It has been shown that the subjective viewing of reconstructed images is often improved if non-linear quantising is employed [10]. This is because the incremental changes in brightness are much less noticeable if the brightness level is high than if it is too low. Hence, the density of quantisation levels should be greater for the lower brightness levels. This is shown in Figure 5.22, where a linear quantisation law (a) is compared with Gaussian quantisation (b). Gaussian quantisation has the property that the probability that a sample will be quantised to a given reconstruction level will be the same for all quantisation levels, assuming that the input is also Gaussian. This has some relevance in digital data transmission, since it implies that a constant word-length code can be used for each quantised sample.

5.4.6 NOISE CONSIDERATIONS

Noise is a limiting factor in digital processing of picture information. This can enter into the digitising-processing-reproduction process in a number of ways.

The subjective effects of noise, no matter how introduced, are intimately associated with resolution, contrast, brightness and local viewing conditions. The effects are quite different, however, for purely machine processing, so that care needs to be taken when assessing the effects of added noise in such operations as contrast expansion or computer enhancement.

Leaving aside the quality of the original source such as the effects of 'grain noise' in a photographic image, we will commence our consideration of noise at the point of sampling and quantisation. The noise problem with quantisation is

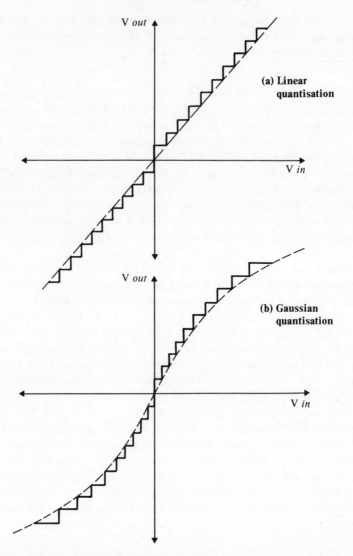

Fig. 5.22   Quantisation of picture information

quite simply that the signal being quantised is actually signal plus random noise so that there is a finite probability that the alternative and incorrect 'next level' to the signal level will be chosen (see Fig. 5.23).

An analysis of the quantisation errors introduced in this way was given in Section 5.2.3. It is interesting to note that whereas with zero noise the quantised value of a given signal will always be the same no matter how often quantisation is attempted, the position with signal plus noise is quite different. Here there is a finite probability that the quantised value will be different for each quantising operation on the signal plus random noise. Consequently, by taking a large

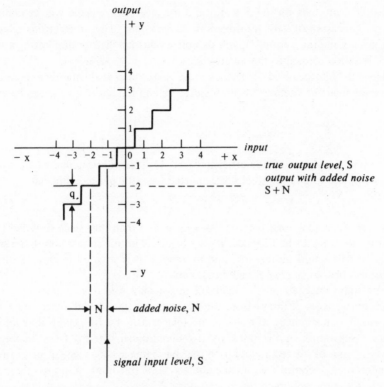

Fig. 5.23   Effect of noise on quantisation

number of samples of the image to be digitised, an average value of the quantised values obtained will yield a truer representation of the image. This is, of course, the principle of image enhancement by averaging, which is in common use.

Choice of quantisation level will be determined by a number of factors. One of these will be cost of processing, since the larger the number of possible grey levels for each picture element then the greater the digital word size and size of the matrix representing the digitised picture. However, we may be concerned with obtaining maximum fidelity in the reconstructed image and this will demand a large number of quantised levels. There is also a limit to which we can take this, and this limit is determined by the average or r.m.s. noise of the available image.

If we define an acceptable system as one in which the quantising noise is equal to the r.m.s. sum of all earlier noise sources in the data acquisition system, then it can be shown that the ratio of the quantising step size to the r.m.s. noise, $B = q/n_{rms}$, reaches an optimum at about $B = 3$. Thus, there is no point in digitising the picture to grey levels finer than this.

What are the sources of noise comprising this critical r.m.s. level? The most significant of these is that arising from the light detector used. In many cases, this is a photoelectric detector or photomultiplier. With such a system, the measured light falls on the detector and the electrons released will be counted to give a measure of light transmission. The number of photoelectrons released in a given sampling period, Q, will exhibit a Gaussian distribution having a standard deviation σ equal to the square root of the average value, $\sqrt{Q}$.

Since the photocathode efficiency must be less than 1, the number of photons is greater than the number of photoelectrons and the photon noise can be given as

$$N_e = \frac{\sigma}{\sqrt{\epsilon}} \qquad (5.13)$$

where ϵ is the photocathode efficiency ($\epsilon < 1$). This will be added to the photo-emission noise to give the total noise emitted from the photocathode as

$$N_t = \sqrt{(N_e \cdot \epsilon)^2 + \sigma^2} = \sigma \sqrt{(1 + \epsilon)} \qquad (5.14)$$

However, the actual noise recorded may be greater than this, since stray light can fall on the image to be digitised. It may be noted in this connection that a small image signal should be less affected by noise if it is situated in an area of low brightness than in an area of high brightness.

The noise imposed on the digitised image may well have a particular noise spectrum which is different from the spectrum of the desired image. As is well known, it is possible to transform the image into the frequency domain and effect a separation of the noise by two-dimensional filtering (see Chapter 3). Consideration of the techniques used to achieve image enhancement of this form by computer operations is outside the scope of this book, although these are, strictly speaking, pre-processing operations.

References

1. SHANNON, C. E. A mathematical theory of communication. *Bell Syst. Tech. J.*, **27**, 1948, 623–56.
2. BRACEWELL, B. *The Fourier Transform and its Applications*. McGraw-Hill, New York, 1965.
3. CHAMPENEY, D. C. *Fourier Transforms and their Physical Applications*. Academic Press, London, 1973.
4. ANDERSON, T. Optimal control logic for successive approximation A–D converters. *Computer Design*, **11**, July 1972, 81–6.
5. YUEN, C. K. Another design of the successive approximation register for A–D converters. *Proc. IEEE*, **67**, 1969, 873–4.
6. YUEN, C. K. Analog-to-Gray code conversion. *IEEE Transactions on Computers*, **27**, 1978, 971–3.
7. HUANG, T. S. *Picture Processing and Digital Filtering* (Ch. 6). Springer-Verlag, Berlin, 1975.
8. ANDREWS, H. C. *Computer Techniques in Image Processing*. Academic Press, New York, 1970.
9. CHIEN, R. T. *et. al.* Hardware for visual image processing. *IEEE Transactions on Circuits and Systems* (CAS 22), **6**, 1975, 541–51.

10. WIDROW, B. A study of rough amplitude quantisation by means of Nyquist sampling theorem. *IRE Transactions on Circuit Theory* (CT-3), 1956, 266–76.

ADDITIONAL REFERENCES

SCHMID, H. *Electronic Analog/Digital Conversions.* Van Nostrand, New York, 1970.

SHEINGOLD, D. H. *Analog–Digital Conversion Handbook,* Analog Devices, Norwood, Mass., 1972.

HOESCHELE, D. F. Jr. *Analog-to-Digital/Digital-to-Analog Conversion Techniques.* Wiley, New York, 1968.

IEEE Transactions on Circuits and Systems (special issue on analog/digital conversion, CAS-25), 7, July 1978.

RABINER, L. R. and GOLD, B. *Theory and Application of Digital Signal Processing.* Prentice-Hall, Englewood Cliffs (NJ., USA), 1975.

Chapter 6
Data Acquisition Systems

6.1. Introduction

In previous chapters, we discussed individual modules contained in data acquisition systems. In order to give the reader a better understanding of how the components described fit together in complete systems, this chapter provides a broad survey of systems that perform some data acquisition function. In addition, some data acquisition components not discussed up to now will be considered here in the systems context.

Although there is a great variety in the characteristics of different data acquisition systems, we are able to group them loosely under the following four categories:

1. Data display systems: systems that measure signals and reproduce the information immediately in a form suitable for human inspection.
2. Data recording systems: systems that record measured signals so that subsequent analysis may be performed on the recorded information, usually at a later time or a different location.
3. Data processing systems: systems that measure signals and process them immediately to produce new information, usually in a form that is much easier to interpret, or with the information greatly reduced in quantity. The former is a data transformation process, and the latter a data reduction process.
4. Integrated data systems: these systems not only process the input signals immediately, but also use the new information directly to perform some control function. That is, they produce output signals which are sent immediately to devices under the control of the data system. The devices may be the same ones that produce the input signals, in which case we call this a feedback data system.

Taking a global point of view, the first three categories may be considered as parts of an integrated data system, since the ultimate purpose of any data acquisition system is to use the data in some human or machine decision-making process which will, in due course, affect some other systems under the control of the decision makers.

However, this is a somewhat philosophical point. It is nevertheless useful to distinguish between these four categories in order to discuss their different methods of operation and equipment requirements.

6.2 Data Display Systems

Data display systems are an essential part of many industrial plants where real-time monitoring of plant activities is required. They are also present in the context of environmental monitoring, patient observation, and in everyday life. For example, the fuel gauge of an automobile is a simple data display system. The original information, the level of petrol in the tank, is measured by some device, say a float, and then transduced into electrical form, and displayed on a meter for examination by the driver to enable him to decide when to fill up the tank. In other words, a data display system consists of, at one end, devices for signal measurement and, at the other end, output devices that reproduce the information for human attention. There are, however, a wide variety of display mechanisms. Alarm bells, for example, are binary devices which are either on or off. Meters, on the other hand, display a continuous range of values at relatively low speed and precision. The most modern equipment tends to have high-precision digital displays. Finally, periodic or slowly varying functions of time may be displayed on an oscilloscope or other device.

Figure 6.1 shows the general configuration of data display systems. Note that, in general, the output of measuring–transducing devices require some treatment, for example amplification, before it can be used to drive display devices. In particular, A–D conversion is necessary before signals can be displayed digitally.

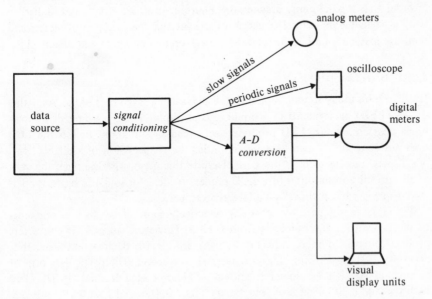

Fig. 6.1   General configuration of data display systems

It is also generally true that data display systems work usually with slowly varying signals. Meter dials have considerable inertia and cannot follow a rapidly varying voltage. While a digital display can change much faster, it would be difficult for the human observer to discern rapidly changing numerals. The same can be said for oscilloscope traces, except when the information displayed is periodic, so that it repeats itself regularly for examination. In most cases, rapidly varying signals either have to be recorded for subsequent processing, or must be processed in real time by a machine matching their speed of variation.

Data measurement, amplification and A–D conversion have been discussed in previous chapters and here we shall be considering data display equipment.

6.2.1 METERS

Out of the great variety of meters three are the most basic: galvanometers, ammeters and voltmeters. The first measures whether electric current is flowing through the meter, the second measures how much, and finally voltmeters measure the voltage difference across a circuit. Voltmeters are applied by connecting their terminals between two points in an electrical circuit. In contrast, to measure the current in a circuit we must break the circuit at some point and insert the ammeter so that the current flowing in the circuit has to go through the meter. Thus, voltmeters are applied in parallel, while ammeters and galvanometers are connected in series. These factors dictate first that an ammeter must have a small impedance compared with the circuit so that its insertion will not disrupt the normal current flow. Otherwise, the process of measuring the current will actually reduce it, making the measurement inaccurate. Secondly, a voltmeter must have a high impedance compared with the circuit it is connected to in order that it would only draw minimal current from the circuit and so disturb it by a negligible amount. Discussion of the internal construction of meters and their use may be found in the references given on measurement equipment [1].

6.2.2 DIGITAL DISPLAYS

It is common today to build numerical displays using seven bars of light-emitting diodes (LED) or segments of a liquid crystal display (LCD), arranged in a particular pattern. Whilst LED produce easily readable luminous figures the LCD does not generate its own light but modifies the transmission of available light. No display can be seen in the dark. Despite this major disadvantage, the very small current consumption of a LCD renders it ideal for long-life battery operated displays, such as those in digital wristwatches.

Either system is used by activating selectively some of the bars or segments to display any of the numerals from 0 to 9. However, since A–D converters produce binary numbers, whereas humans work with decimal numbers, it is necessary first to perform binary-to-decimal conversion. The particular form of decimal conversion employed is known as **binary-coded-decimal** (BCD). Here each decimal digit consists of four binary bits, 0000 to 1010, so that it is necessary to derive seven bits from them by means of a Boolean logical network in

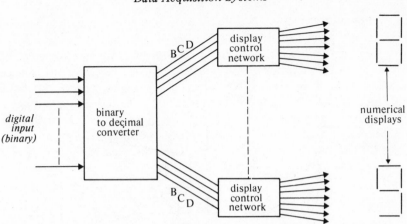

Fig. 6.2   Connecting a digital display

order to control the seven bars. Thus, a complete digital display is seen to consist of three parts, shown in Figure 6.2: a BCD converter, a Boolean control
network and the numerical display unit. It may also be found necessary to add
more circuits to handle signed numbers, or to make the display more general
purpose, e.g. to be able to display the same number in other representations
such as octal or hexadecimal, which will require different converter logic. Values
that have a wide range have to be displayed in floating point format, with a
mantissa and an exponent for each.

Digital displays can be purchased readily from component manufacturers and
a large range of specifications are available. It is also possible to acquire complete
systems of digital display voltmeters or multi-range meters, and we shall discuss
these briefly in the next section.

In digital systems, data is displayed visually by means of a **visual display unit**
(VDU). This incorporates a cathode-ray tube which displays alphanumeric information line-by-line across the tube face similar to a television display (in fact
VDUs are often described incorrectly as television display units). The necessary
logic to enable the incoming data stream to be displayed in this way is incorporated within the VDU which often carries a keyboard for manual insertion of
data by the operator. Keyboard VDUs are commonly used as the manual input
and display device for a digital computer or microprocessor.

6.2.3 OSCILLOSCOPES

An oscilloscope consists of an electron beam emitter that directs a narrow beam
on to a phosphor-coated screen to produce a small light spot. The co-ordinate
position of this light spot on the screen is controlled by applying electrical poten-
tials across two pairs of electrodes, arranged at right angles, through which the
beam passes; this is illustrated in Figure 6.3. The oscilloscope may be used to display graphically on the screen a time-varying function such as that shown in
Figure 6.4. We can see from this that the value of the function y is used to

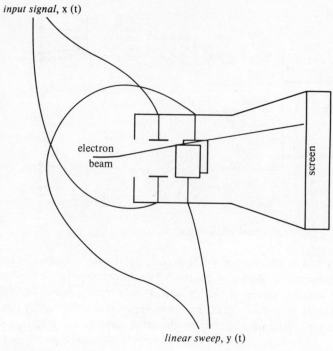

Fig. 6.3   Oscilloscope

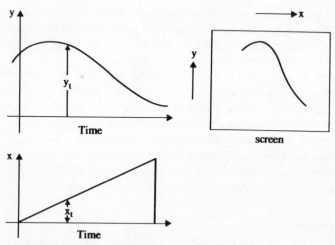

Fig. 6.4   Making an oscilloscope trace

control the vertical position of the beam. As the function changes, the light spot moves up or down. At the same time, a linearly increasing voltage function, x(t), is applied to the x-electrodes and causes the light spot to move horizontally across the screen at a constant speed. This provides an X–Y plot of the function on the screen as the light spot moves across it. Selection of which part of the function to inspect can be arranged by timing the start of the horizontal sweep and by varying the rate of the sweep. If the sweep is fast, only a small part of the function will be displayed on the screen. Because oscilloscopes can be made to store the data on the screen almost indefinitely, it is possible to select only a small part of a rapidly varying function for extended study. However, more usually the oscilloscope is used to display functions which are periodic, at least approximately. As shown in Figure 6.5, if we make the sweep duration have the same period as the function, y(t), then each sweep will cause the trace to follow exactly the route taken by the preceding sweep. We have then obtained a stationary display of a periodic function, but to do this successfully, one has to know the period of the function exactly in order to initiate each sweep at the right time. Since the period is usually unknown until after we have examined the function, some method is required to overcome this problem. A common technique is to **trigger** the sweep by using the signal input function; the oscilloscope is set to start its sweep whenever it detects that the function has just attained a specified value. As can be seen from Figure 6.5, because the function is periodic, the start of the sweep always occurs at the same point within each period, known as the trigger point. Figure 6.6 shows how one can study the function for part of the period only, by delaying the application of this trigger pulse. Here, a pre-sweep waveform generator is initiated when the signal amplitude reaches the trigger level as described above. This waveform is itself associated with an adjustable threshold level, so that the actual screen sweep waveform does not commence until the threshold level is reached and terminates at the triggering point of the signal. This somewhat complex method of delayed triggering is used due to its reliability and the accurate control of trigger delay obtained.

6.3   Data Recording Systems

We can produce data recording systems by making some relatively minor extensions to data display systems. As shown in Chapter 4, chart recorders were initially just meters having a pen in place of the needle. Similarly, if we replace a digital display by a printing device, we would obtain a 'hardcopy' record of the numbers. Many oscilloscopes today have optional hardcopy units which may be used to record the trace observed on the screen on to a paper record. Figure 6.7 shows the general configuration of data recording systems.

Despite this apparent simplicity, data recording systems can be quite elaborate. A number of digital voltmeters or multi-meters are available, which are in fact fairly complete data recording systems. They have widely variable input impedances, so that they may be used in a variety of measurement tasks at different ranges and precision. They may also include pre-processors that can perform signal conditioning or a.c.–d.c. conversion. The output may be a CRT display or hardcopy and the necessary equipment for computer interfacing or

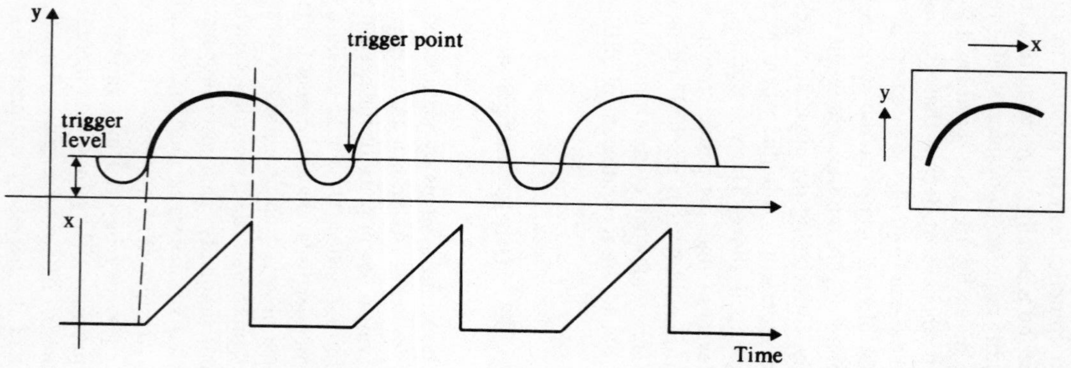

Fig. 6.5 Capturing a periodic function

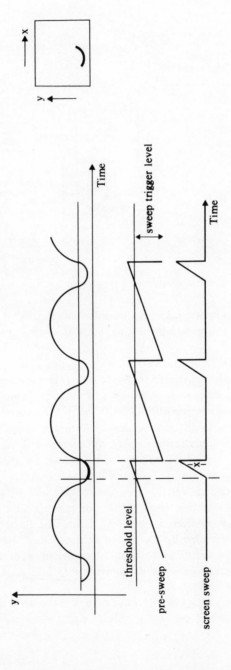

Fig. 6.6 The delayed trigger process

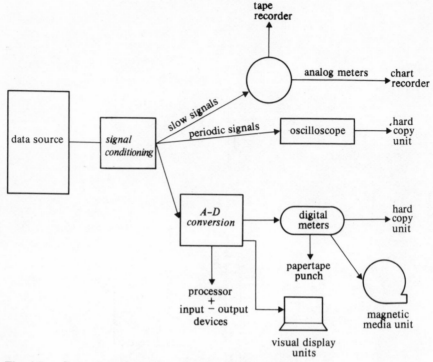

Fig. 6.7   General configuration of data storage systems

magnetic tape recording may also be included. Finally, it may be possible to carry out some minor post-processing, such as scaling or calibration with the equipment.

An example of a sophisticated device of this kind is the Bryans BS8000 'intelligent' recorder illustrated in Figure 6.8. This combines the operation of an eight-channel recorder, an x–y plotter and a digital voltmeter. Program control of these devices is obtained through the use of a microprocessor with which is associated a magnetic disc storage device. Although accepting analog inputs and producing a continuous analog display, the operation of the device is carried out entirely digitally using the keyboard entry shown for editing, storage, mathematical processing and output operations.

Typical applications for this recorder include direct comparison of experimental results, calibration from stored calibration curves and examination of data for trends and other programmed tasks. The flexibility of this type of recorder is made possible by the microprocessor, which we consider in the next chapter and in the ability to store quite considerable numbers of digital samples on magnetic disc (150 000 points equivalent to 300 graphs in the Bryans recorder).

Sampling rates are limited with this type of recorder to the audio range up to about 20 000 samples per second. Much higher rates are possible if the data

Graphical display
Ordinary or graph paper can
be used; axes are
automatically drawn.
Electrostatic hold-down is
convenient and trouble free.

Numerical display
Digital display provides
monitoring and simplifies
operation.

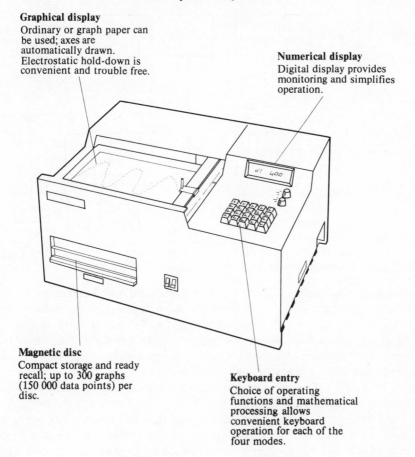

Magnetic disc
Compact storage and ready
recall; up to 300 graphs
(150 000 data points) per
disc.

Keyboard entry
Choice of operating
functions and mathematical
processing allows
convenient keyboard
operation for each of the
four modes.

Fig. 6.8   Bryans 'intelligent' recorder (included by courtesy of Bryans Southern
Instruments Ltd)

acquisition is limited to a shorter time period, since the problem is one of storage
for the number of samples obtained, rather than speed or duration of the data.

These faster devices are known as **transient recorders** and are valuable for the
study of short-lived phenomena, such as, for example, disturbances induced in
electric power supply lines.

A transient recorder is, in effect, a time record translater which records the
short-lived transient and permits its reproduction on a longer time scale so that
its form can be studied. Temporary storage of the transient information is required and this is usually carried out digitally. Various forms of analog recorder
and reproduction have been attempted, such as photography of oscilloscope
traces and the use of storage oscilloscopes, but these generally lack resolution
and flexibility.

A schematic diagram of a digital transient recorder is shown in Figure 6.9.
The principal components are an A–D converter, a digital memory and a D–A

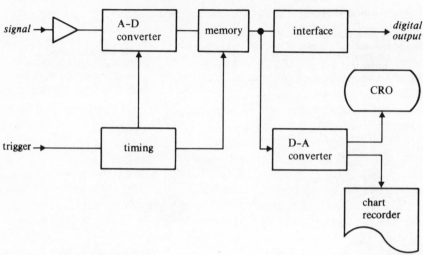

Fig. 6.9   A transient recorder

converter. The resolution of the recorded transient will be affected by the sampling rate used in the A–D converter, since this controls the highest effective frequency that can be recognised (see Ch. 5). However, the size of the digital memory is fixed, so that in order to make full use of this, the sample length (i.e. number of samples taken) remains constant with the sampling speed adjusted to correspond with the different time periods of recording that may be required. The effect on resolution for the recorded transient must thus be taken into account.

The transient recorder may be put into operation by external triggering or use made of the onset of the transient itself to initiate the recording process. In the latter case, it will be necessary to introduce a delay in the transient signal reaching the recorder, so that the beginning of the transient will be recorded. An alternative arrangement is to permit continuous A–D conversion and recording to take place where the memory behaves as a dynamic shift register which, when full, will discard the oldest sample stored to make room for a new sample. Sampling can then be made to cease at a preset number of samples (the memory size) after the trigger signal has been received.

After the transient has been recorded, the contents of the digital memory are passed through the D–A converter in the same order in which they were stored and the analog signal is displayed on a suitable device, e.g. a CRT screen, or a pen recording is made. Since the speed of read-out from the memory can be controlled to suit the reproducing device, a time translation may be effected enabling a fast transient to be displayed on a much longer time scale than the original phenomenon, thus permitting detailed study of it to be carried out.

6.3.1   ANALOG VERSUS DIGITAL RECORDING

In most cases, the right recording medium for a particular system is determined naturally by the circumstances in which it is derived. This will depend, for example, on the amount of data, the speed of the signal, the system configuration and the use we wish to make of the data [2]. For example, chart recordings made in real-time are useful only for relatively slowly varying data. Under some circumstances, chart recordings may be manually processed to extract the useful information in the form of a continuous electrical signal, as described in Chapter 4. When the amount of data is large but the processing involves only an examination of occasional significant features, such as isolated peaks, it is possible to extract the information by rapid manual inspection of a long roll of recording and translation of only the relevant sections of the data. For many purposes, chart recordings are regarded as useful only for providing a back-up record such that, if any of the data are lost, the information can, in principle, be recovered, although at the cost of some painstaking labour.

To take a second example, a hardcopy unit or papertape punch attached to a digital voltmeter can constitute a useful technique for recording small amounts of data. The former has the advantage that the information can be inspected visually, whilst the latter is in machine-readable form and may be fed into a computer directly using a tape reader, thus avoiding key-punching operations. Similar comments apply to other magnetic media, such as magnetic tapes. On the other hand, if the digital voltmeter is on-line to a computer, it proves more convenient to inspect the data through the use of an editing program at a terminal linked to the computer, retaining the hardcopy or papertape output as a back-up record.

Because of the determining effect of actual circumstances, the choice of the best recording medium often involves a decision between analog and digital systems. In the remaining paragraphs, we make some comments on this question.

Generally speaking, analog recording permits a higher data density. For example, most analog magnetic tapes permit simultaneous recording of several channels, whereas a digital magnetic tape unit has to use the whole width of the tape to record just one number, since a number consists of several bits, each of which has to be represented by a magnetic dot. Analog tapes also have higher densities in the longitudinal direction, since with digital tapes the two rows of magnetic dots for consecutive numbers must be well separated. Thus, when very large amounts of data are involved it is usually more economical to record them in analog form.

The main advantage offered by digital recording is higher accuracy and easier handling of the recorded data. Although both analog and digital recording devices suffer from such problems as tape imperfection, grain noise, uneven tape speed, external interference, etc., these do not affect the two kinds of data in the same way. An error caused by the analog system has an immediate effect on the analog signal recorded, whereas digital signals can only change in discrete steps so that the noise must be comparable to the size of a step before the effects of this are noticeable in the recording. A digital value consisting of, say, n pulses of voltage $\pm V$, each representing a binary digit, can be corrupted only if a noise pulse of magnitude $\geqslant V$ is superimposed on one of the digits. Furthermore, by suitable shaping of the pulses and the addition of error production information (see Ch. 8), it is possible to detect, or even correct, errors in digital information, provided the noise power is below some specified amount.

Of course, the very process of digitisation will introduce sampling and quantisation errors, as discussed in Chapter 5. However, such errors are both estimable and controllable. A mathematical framework exists permitting the calculation of the amount of error arising from sampling and quantisation, so that we can adjust these errors according to our requirements by changing the sampling rate, the number of digits, etc., to ensure that errors remain within tolerable limits. In short, in digital recording, errors are deliberately introduced to prevent or minimise errors of other kinds.

A further advantage of digital recording is the relative ease with which data may be converted from one form to another, for example from papertape to magnetic tape, or from disc files to printed output. This can allow the multiple use of the same data for several different purposes.

6.4 Data Processing Systems

Few current systems are pure data recording systems in the sense that the raw data produced by the signal source are simply recorded. More commonly, some processing is carried out to generate new data, which are then recorded. However, it is necessary to distinguish between two kinds of processing: pre-processing, the purpose of which is to extract existing signals from the raw data by eliminating noise, calibration errors, etc., and the data processing itself. Whereas the need for pre-processing is determined mainly by environmental factors, such as the presence of external interference and the behaviour of instruments used, the nature of the problem itself determines the kind of data processing we wish to carry out. Depending on the problem at hand, hardware or software processing may be called for in varying degrees of complexity. In the present section we shall discuss a number of common data processing systems.

6.4.1 THRESHOLD DETECTION

Detecting that the input signal exceeds some specified value and only transmitting data under this condition may be considered as a simple and also very common data processing operation. The process consists of the conversion of an analog input signal into a digital output (1 = value exceeded; 0 = not exceeded). The operation may be carried out conveniently using an operational amplifier which has a high amplification factor. This is shown in Figure 6.10. Given an ideal amplifier, the output y would be

$$y = A(x - v) \tag{6.1}$$

where v is the threshold value and A the amplification factor. If, for example, $A \simeq 10^5$ and $x - v = 10^3$, then $y = 100$. In actual fact, the amplifier can only deliver and output a value within the range of, say, $\pm V$, and whenever $A(x - v) > V$, the output is simply V, and $A(x - v) < -V$ would cause the output to

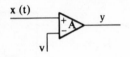

Fig. 6.10   A threshold detector

be − V. Thus, the input–output relation has the form shown in Figure 6.11. We see that, because of *saturation effects* the output is, for all practical purposes, binary: it remains at −V if x is only slightly below v, and + V if it is slightly above. The only two possible output voltages + V and −V may thus represent digital 1 and 0, respectively.

By adjusting the amount of feedback, the amplification factor A may be varied, as described in Chapter 3. If A is 10^5 and V is 5 (fairly typical values), then x rising from $v − 0.00005$ to $v + 0.00005$ will cause the output to swing from 0 to 1. For many applications this may be too sensitive. For example, if x is contaminated with noise, then we would not want y to change from 0 to 1 just because the input momentarily exceeds v owing to noise. Hence, the quantity V/A, which indicates the sensitivity, as well as the input noise tolerance of the threshold detector, should be chosen carefully depending on conditions. A large A increases sensitivity, but it also requires that the input has a high signal-to-noise ratio, so that additional signal filtering may become necessary. Very high amplifications may require several amplifiers connected in series, although such systems will need to be carefully designed to avoid noise and instability problems. Two simple extensions of threshold detection are given below.

1.   Alarm recording: The threshold detector discussed earlier gives as its output a 1 only when the input actually exceeds the threshold, since it returns to 0 as soon as the input falls back below the threshold value. For some systems, it may be necessary to cause the alarm to stay on until some action has been

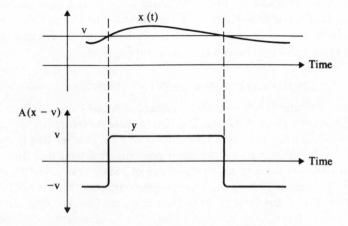

Fig. 6.11   The operation of a threshold detector

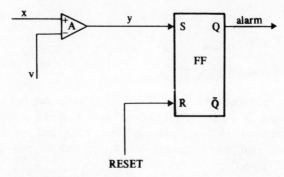

Fig. 6.12   An alarm recording unit

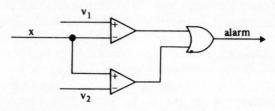

Fig. 6.13   Range detection unit

taken, following which the alarm may be turned off manually. This can be achieved by adding a monostable circuit or a flip-flop to the system as shown in Figure 6.12. We see from Figure 6.12 that if input x exceeds v for a short period, y would be momentarily 1, which makes $S = 1$ and $R = 0$, so that Q becomes 1. It will remain 1 even after y has gone back to 0 (see Section 1.6.1). The only way to make Q become 0 again is by manually switching the RESET input to 1, after the alarm has been handled.

2.   Range detection: In some systems we may wish to ensure that x remains within the limits of v_1 to v_2. An alarm is raised if x goes outside these bounds. The circuit in Figure 6.13 achieves this. The two amplifiers produce the information $x > v_2$ and $x < v_1$, respectively, and the OR gate output is 1 if either input is 1, i.e. if x falls outside either limit.

6.4.2   DURATION DETECTION AND EVENT COUNTING

For some applications it is necessary to record how many times or for how long a signal exceeds a given threshold. The first of these is determined by the use of an edge-triggered counter. This device stores an n-bit number that is caused to increase by 1 whenever its control input pulse makes a transition from 0 to 1. Figure 6.14 shows the internal configuration of such a device, and Figure 6.15 explains its operation. We see that whenever the output of the threshold detector jumps from 0 to 1, the value in the counter increases. Consequently, starting the system with 0 shown in the counter display, the number of occasions on which x exceeds the value v will form the final number to be recorded in the counter.

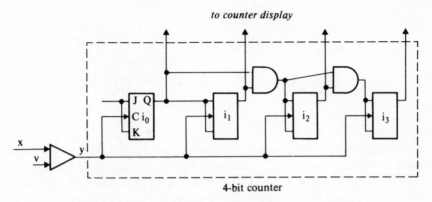

Fig. 6.14   An event counter, with four JK flip-flops

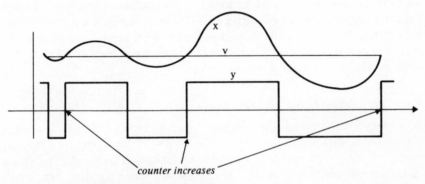

Fig. 6.15   Operation of an event counter

Alternatively, one may wish to record the total duration in which $x > v$. This can be done in several different ways. One is to integrate the output of the threshold detector

$$I = \int \{y(t) + v\}\, dt \qquad (6.2)$$

Since $(y + v)$ is $2v$ if $x > v$, and is equal to 0 if $x < v$, we know

$$I = 2vT \qquad (6.3)$$

where $T =$ duration in which $x > v$. Thus, we can measure the duration by digitising the output of the integrator and dividing the result by 2.

A different method is shown in Figure 6.16. Here, the output of the threshold detector is used to control a sequence of equally spaced clock pulses. Whenever $y = 1$, the pulses are permitted to reach the counter which records a pulse count number, increasing as time proceeds. The longer x exceeds v, the more the pulse count increases, so that the final count registered is proportional to the total duration during which $x > v$.

In order to assess the effectiveness of the two methods for duration detection, a comparison of their respective merits can be made which has a similarity to

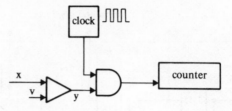

Fig. 6.16   A duration detector

those described for analog-versus-digital recording methods. The first, the analog method, has higher precision, in the sense that if everything worked ideally then it would be the more accurate. This is because, in the second method, the counter increases at certain fixed points of time only, so that shorter lengths of time during which x exceeds v are not recorded at all. However, the analog method is affected by instrumental imperfections in the integrator, particularly through capacitor leakage, whereas the digital counting method would not be subject to such effects. In this sense, the second method is more accurate. Note, however, that the clock must operate at a rate much faster than x is expected to vary, in order to achieve reasonable precision.

6.4.3   STATISTICAL QUANTITIES

It is often a useful exercise to compute the mean, variance and correlations of the measured data soon after acquisition, because of the significant effect the results of these tests may have on methods adopted for data reduction. These quantities summarise a great deal of information about the data, but require much less data storage space than the original data. In many problems, further processing can be performed on the average quantities alone without having to retain the original data, thus saving considerable storage space.

Where a digital computer is used in data acquisition it is a simple matter to write the necessary programs to perform the statistical computation. The mathematical aspects may be found in our companion volume *Digital Methods for Signal Analysis* [3, 4]. In this subsection, we shall discuss briefly analog techniques for such statistical computations.

The mean
The mean, $\bar{x}$, of a function over a period of measurement is

$$\bar{x} = \frac{1}{T} \int_0^T x(t)\,dt \tag{6.4}$$

where x(t) is the measured signal function. Thus, the mean may be measured by the use of an integrator with the product RC set at a value T.

The variance
The variance $V(x)$ is measured as

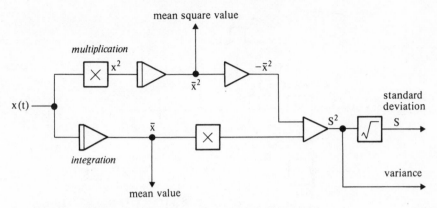

Fig. 6.17   Analog circuit for computing statistical quantities

$$V(x) = x^2 - (\bar{x})^2 = \frac{1}{T} \int_0^T \{x(t)\}^2 \, dt - (\bar{x})^2 \tag{6.5}$$

This shows that we require two integrators, for $x(t)$ and $x(t)^2$, respectively, two multipliers, to compute x^2 and $(\bar{x})^2$, and a subtractor. The standard deviation may be produced from $V(x)$ by taking the square root, which is unfortunately a difficult operation to perform accurately by analog means. A composite analog circuit is shown in Figure 6.17 which produces all these quantities from $x(t)$.

Correlation coefficient
The coefficient of correlation between two functions, $x(t)$ and $y(t)$, is measured as

$$\rho_{xy} = \left\{ \frac{1}{T} \int_0^T x(t)y(t) \, dt - \bar{x}\bar{y} \right\} \Big/ (\sigma_x \sigma_y) \tag{6.6}$$

where σ stands for standard deviation. One can see that this is rather complicated, requiring no less than five integrators to realise the function.

Correlation functions
The autocorrelation function of $x(t)$ is

$$R_{xx}(\tau) = \frac{1}{T} \int_0^T x(t)x(t + \tau) \, dt \tag{6.7}$$

while the cross correlation between x and y is

$$R_{xy}(\tau) = \frac{1}{T} \int_0^T x(t)y(t + \tau) \, dt \tag{6.8}$$

To compute these by analog methods means requires delaying a function by τ, multiplying and integration. Since, however, τ is variable, the whole function has to be stored for repeated evaluation for different values of τ.

A schematic diagram of an analog circuit used for this purpose is shown in

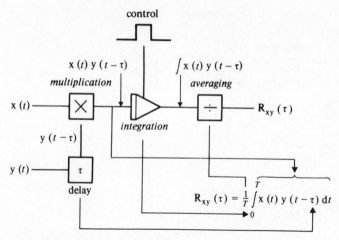

Fig. 6.18   Analog circuits for computing R_{xy}

Figure 6.18. Note that the estimated value of $R_{xy}(\tau)$ obtained here uses $y(t - \tau)$ and not $y(t + \tau)$. However, the results are identical for stationary signals. Owing to difficulties in storage of the function values, it is more usual today to find digital circuits used for the calculation of correlation functions.

6.4.4 SPECTRAL ANALYSIS

Spectral analysis attempts to detect the underlying periodic structures of a function $x(t)$ by computing its power spectrum. This shows the average distribution of the energy in $x(t)$ among different frequency components. One can also study the similarities and differences between the frequency components of two functions by computing their cross-spectrum. Once again, it is more common to employ digital methods for such analysis, though analog techniques, by the use of a bank of narrow band filters, as discussed in Chapter 3, are still sometimes found.

Regardless of whether one adopts a software or hardware system for spectral analysis, a number of parameters affect the overall cost, either in the form of hardware components or program execution cost. First, we need to choose the range of frequencies to be studied; this is known as the **analysis bandwidth**. The second is the **resolution** required, i.e. the narrowness of the sub-intervals into which the frequency range is divided for separate analysis. The total number of output values constituting the spectral estimate is seen to be the bandwidth divided by the resolution. Each spectral value measures the amount of signal power contained in a sub-interval.

It is necessary to make a remark about the effect of another parameter: the centre frequency of the analysis bandwidth. If a signal has an interesting frequency range going from, say, f_a to f_b, then the analysis bandwidth is $(f_b - f_a)$, which means the centre frequency is $\frac{1}{2}(f_a + f_b)$. Generally speaking, the centre frequency has only a minor effect on the complexity of spectral analysis. It is fairly easy to transform the signal such that its spectral structure remains the same except that the whole frequency interval (f_a to f_b) is shifted to (0, $f_b - f_a$). That is, we can easily change the centre frequency without changing the bandwidth and without affecting the validity of the results. In contrast, one cannot reduce the bandwidth without suffering some loss of spectral information.

6.5   Integrated Data Systems

A typical integrated data acquisition system has a configuration similar to that shown in Figure 6.19. Here, we have a data processing system in which the results of processing are also fed back to the source of the data to control its behaviour [5–11].

The design of an integrated data system is usually much more difficult than that of display, recording and processing systems. This is not due only to increased system size. The additional complicating factor is the presence of feedback, in that the present behaviour of the system, as measured by the data acquisition unit, is being used to control its future behaviour. A system with feedback tends to exhibit a more complex dynamic behaviour because this varies with its past history (which is itself affected by factors existing in the further past) and the degree and speed at which past behaviour feeds into future behaviour, as well as by external stimulus.

Here, we take two simple examples: first, consider an automatic street light control system, which turns on the lights when the lightmeter indicates that illumination has fallen below some threshold value and off when it is above this

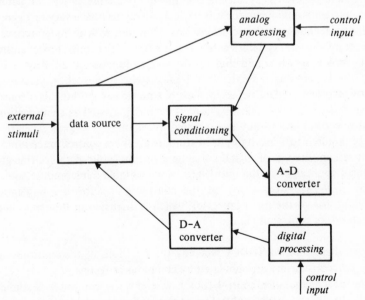

Fig. 6.19   General configuration of integrated data systems

value. Such a system will fail, because turning on the lights immediately increases the illumination past the threshold value, causing the lights to go off, whereupon illumination falls below the threshold and the lights are turned on again. Thus, the system behaves in an oscillatory manner because of the interaction provided through the lightmeter and controller. In order to produce stable behaviour, it is necessary to adopt a more complex system, having two threshold values. In this system, lights are not turned on until the illumination has fallen below a given lower limit and are turned off only when an upper limit has been exceeded. A minimum separation between the two threshold values is necessary such that turning on the lights would not increase illumination from below the lower limit to above the upper limit.

Our other example is a slow-responding thermostat arranged to turn on a boiler when the temperature falls below a threshold of $T°$. However, the thermostat may be such that it does not respond until some time after the threshold has been crossed. By the time the boiler is turned on, the temperature may already be considerably below $T°$. Conversely, the boiler may remain switched on until the temperature is much above $T°$. So here too we have an oscillatory condition caused by feedback, or more exactly by inertia in the thermostat operation. The remedy is obviously to increase the response speed of the thermostat.

These examples illustrate the need to design carefully the amount and the rate of feedback, in addition to considering the amount of data to be acquired, the capacity of the data acquisition equipment, the actual mechanism for converting the output of the data system back into a form suitable for control, and so on.

A second important consideration with integrated data systems is that of reliability and back-up [12]. In the case of the three other categories, failure of the data system would mean the loss of data, which may have varying degrees of seriousness depending on the actual context. However, with an integrated system, the failure of the data system would mean the loss of the control mechanism for the data source, which would thus not be able to function at all. Any malfunction of the system has to be detected immediately, since such a system would send out incorrect control signals, and cause a previously working data source to misbehave. This is why high reliability is required, as well as some method of alternative control procedures on equipment (back-up).

These requirements inevitably lead to more complex control procedures and, for this reason, the use of digital computers for integrated data systems gained widespread application in the mid-1960s, when suitable minicomputers became available [13–15]. Now many of the control procedures are implemented through the use of the microprocessor, which is discussed in the next chapter. Two methods of control are in use:

1. **direct digital control** (DDC), whereby the functions of the control loops at the process interface are simulated by computer programs
2. **supervisory computer control** (SCC), whereby the computer is connected with the discrete analog controllers at the processor interface.

In both methods, the computer or processor implements a method of control by program algorithms which utilise measured process variables. The advent of the microprocessor does provide a cost-effective DDC replacement for the analog controller, but the external connections and adjustments are likely to remain unchanged, since these are determined by the requirements of the control needed in the integrated data system.

We have only been able to make here some general remarks concerning the requirements of feedback control and reliability design. Detailed discussions of these topics may be found in specialised texts, some of which are given in the references at the end of this chapter.

6.6   Interconnections and Standards

Mention has been made earlier of 'intelligent' instruments and the use of the microprocessor to control measuring equipment. The whole of Chapter 7 will be devoted to microprocessors and their uses, but it is worth considering at this point their impact on systems development. A major change that has been brought about in the design of measuring instruments is the acceptance of the General Purpose Interface Bus, or GPIB. This is also referred to as the IEEE-448 Programmable Instrument Interface, an American standard which has also been adopted as a standard by the IEC. It is a general purpose digital interface designed specifically to meet the needs of automatic measurement and test systems. The bus provides a relatively simple way to interconnect up to 16 programmable instruments, i.e. instruments that have the appropriate digital connections available. A wide variety of these is now available and includes analog or digital meters, recording devices, signal generators, timers and counters, A–D and D–A converters and, of course, microcomputers. Strictly speaking, the GPIB system is simply a multicore cable connecting the various pieces of equipment, although it will usually be necessary to associate this with a specified controller which will contain a microprocessor. For the moment, however, we will only consider the cable connections and the function they carry out.

Figure 6.20 gives the structure of the general purpose interface bus. There are 16 lines in the bus. Eight lines comprise the data bus D1–D8; these are used to transfer data one byte at a time from one interface device to another. They are also used to transfer address information, as we shall see. Three lines are designated as control or 'handshake' lines. These control how information is transferred from one interface device to another. This handshake operation is based on an interlocking relationship between a transmitting device and one or more receiving devices. Briefly, one line indicates that data is available for transmission, a second indicates that the receiving device is ready to accept it and the third line informs the transmitting device that the byte of information has been accepted. A correct sequence of signals on these three lines enables information to be transferred over the data bus, one byte at a time.

The remaining five lines are the general interface management lines. These are

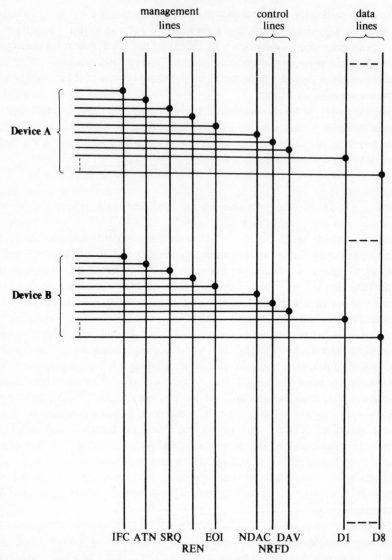

Fig. 6.20   General purpose interface bus structure (GPIB)

used to co-ordinate the flow of information on the 16-line bus as a whole. Referring to Figure 6.20, these lines operate as follows:

1. The Attention Line (ATN) is driven by the controller. When this is set at a
    logical 1, the data on the data bus is interface control information; if it is set
    at a logical 0, then the data bus contains device-dependent data. In the first
    case, for example, the information may be address data and all of the con-
    nected devices decode this to determine which of them is being addressed.

On the other hand, if ATN is at logical 0, the information may be programming data to set one of the inputs of the device for a particular purpose.

2. The Interface Clear Line (IFC) is also driven by the controller and used to clear the complete interface system and reset it to its idle state.

3. The Service Request Line (SRQ) is used by a connected device to inform the controller that the device requires attention. The Remote Enable Line (REN) is driven by the controller to indicate whether the devices are to be controlled by means of signals along the interface or from their own internal controls. The End of Identity Line (EOI) is used by a transmitter to indicate that the end of a multiple-byte transfer sequence has been reached.

The basic operation of the bus is to transfer data between a specified transmitter in a device to one or more receiving circuits in another specified device. The transfer is set up and supervised by the GPIB controller and data bytes are transferred in a byte-serial bit-parallel manner down the eight data bus lines. Each transmitter and receiver module in the connected devices has an address set by switches on the rear panel of the device. An address module can recognise this by address information being transmitted over the data bus in conjunction with the IFC line. It is then ready to receive a stream of data byte transfers without further intervention from the controller until an EOI signal is received.

An example of the way in which this bus system may be used is shown in Figure 6.21. The frequency and phase response of a unit under test is required. The input to the unit is from a programmable signal generator which is capable of providing an alternating input of variable voltage and frequency. The output is monitored by a digital voltmeter set to read alternating voltage, and applied to a digital frequency—phase meter. A permanent record of the output could also be obtained by connecting a printer or plotter to the system, the only requirement being that it too must have the appropriate GPIB connection. The sequence of testing which involves varying the frequency generated, the range of the measuring equipment and form of the printed output is all controlled by a microprocessor pre-programmed with the appropriate instructions. Each item of equipment connected to the interface bus has interface modules, enabling it to act as a receiver, a transmitter or a controller. The basic sequence of operations in the bus is for the controller (the microprocessor) to specify the transmission device (the signal generator) and one or more of the receiving devices (voltmeter and plotter, etc.). The initial measurement step with this system commences with the microprocessor's controller module taking possession of the bus. This then sets up a data transfer by which the microprocessor can select the output frequency of the signal generator. Other devices may be addressed and initialised in turn, when the controller module will send a trigger message down the bus putting the measurement into effect. After the measurement has been taken, the controller will arrange for the readings of the various instruments to be transferred to its memory for analysis and storage and to the output devices (printer and/or plotter). To do this, it specifies first the voltmeter as the currently active transmitter and then the mircroprocessor and printer

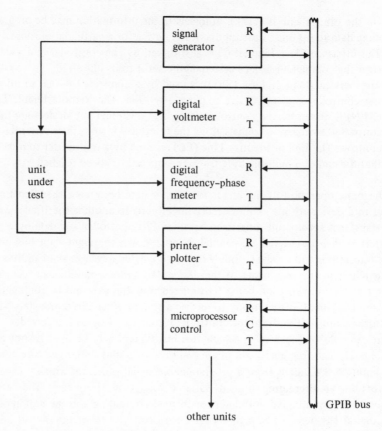

Fig. 6.21   Use of GPIB for unit testing

as the currently active receivers. When this data transfer is triggered, it proceeds automatically, even though the rates at which the printer and microprocessor can accept data are widely different. When the transfer is complete, the frequency–phase meter is specified as the transmitter and the frequency and phase readings similarly transferred. The microprocessor then proceeds to take the next measurement, continuing until the whole test routine is completed and a full frequency–phase response has been taken.

The use of the GPIB to link the various pieces of test equipment to the microprocessor thus allows the complete measurement routines required to be programmed, thus allowing the use of sophisticated instrumentation by unskilled operators. With a suitable change in the microprocessor program, it would also be possible for automatic calibration to be carried out and the necessary documentation information associated with the hardcopy print-out of the tests. We will be considering the microprocessor and programming implications of this type of operation later in Chapter 7.

6.7   Examples of Data Acquisition Systems

So far, we have not shown any examples of data acquisition systems. This is because actual data systems, except very simple examples, cannot be classified neatly into the four categories we have given. Every system has some data display facility to permit an operator to at least monitor the system and occasionally inspect the data. The system will also have some recording facilities to provide back-up contingency against data loss or, at the very least, a summary of activities and will carry out some sort of data processing as well as some measure of instrumental control. Thus we may only use our categorisation to provide a framework for discussing the various functions involved. This is why we give examples here in a separate section, rather than under the previous headings.

Figure 6.22 shows the block diagram of a system for the on-line processing of electro-cardiographs. The system controls multiple measuring devices (only one is shown in the diagram) consisting of individual patient consoles and chart recorders. The signals derived from the patient are selectively analysed through the analog selection unit since there is only one A–D converter. The digital data are saved on magnetic tapes. Two additional computer I–O units, a teleprinter console and a papertape reader and punch, are required to provide operator access to the data and for program input.

The above system handles only one kind of input signal. Figure 6.23, on the

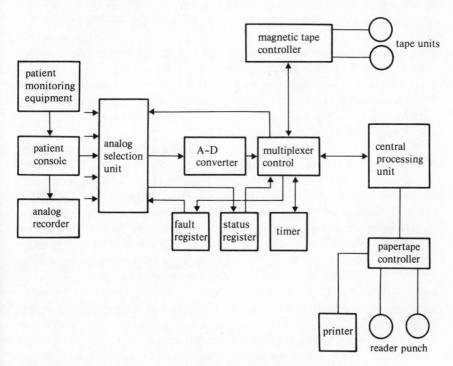

Fig. 6.22   An on-line electro-cardiograph measurement system

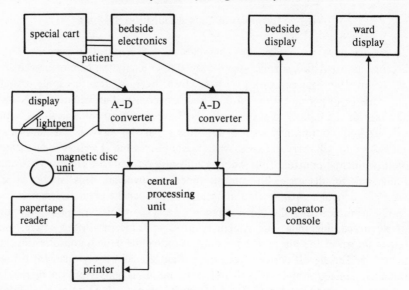

Fig. 6.23   An integrated hospital patient data acquisition system

other hand, shows a larger, more integrated system designed to collect a variety of input data. Analog inputs from a patient are measured and displayed by bedside electronic instruments and sent to the A—D converters. Attached to the conversion equipment is a central display oscilloscope for selective examination of input from the various sources. Bedside keyboards are provided to permit the input of digital data concerning the patients nearby, and there are visual display units so that information about a patient may be fetched from the central computer store and displayed for the medical attendants at the bedside. Because of the large scale of the system, the processor needs to be fairly powerful, and large-capacity magnetic discs are required to store the data and programs and to provide rapid on-line access.

Figure 6.24 shows an interactive image processing system. The picture scanner on the left digitises a picture for input into the machine via the A—D converter. The scanner output consists of the grey levels measured as a series of pixels distributed over the picture area. The actual scanning process, whether a raster pattern or some other pattern as required by the particular problem, is under the control of the computer connected to the A—D device, so that information relating to a particular pixel with its x—y co-ordinates will be automatically available within the computer. All this information is subject to processing in the computer, and digital images produced from it are sent to an image storage unit for display on the TV monitor screen. While it is possible to store the values representing the picture in the computer itself, the data storage requirement will be quite large, and high-speed I—O channels will be needed to display the picture. The use of a separate image storage unit for this purpose relieves the load on the computer. The operator at the monitor examines the output and initiates program commands for further processing or fresh scanning through the monitor

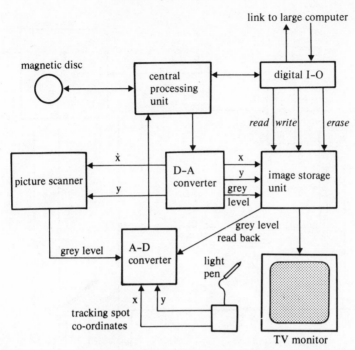

Fig. 6.24   An interactive image processing system

keyboard. As well as using the keyboard, the operator is able to control the processing operators through the use of a **light pen**. This is a small light-sensitive device which is used to point to selected areas of the display which contain features or displayed commands related to the desired processing operations. Light information detected by the light pen is transferred to the computer. Following selection in this way, the X–Y tracking co-ordinates are identified and caused to effect the program contained in the processor so that the required action is taken. Many image processing systems are also interfaced to a larger central computer, to which whole pictures or parts may be sent for processing operations that require greater computing power and storage space than that available on the local processor.

Figure 6.25 shows a speech processing system. The speech input is measured by the microphone, amplified and subjected to several different bandpass filters. The filter outputs are averaged and multiplexed before being digitised through a common A–D converter. The computer will later separate the various data streams and process them, with each stream providing different spectral information. The data acquisition process is controlled by software through a standard interface, as described in Section 6.6, with the program specifying the channel to be selected, whether artificial noise is added to the input, and whether a logarithmic function is to be imposed on the composite input channel.

Our final example is given in Figure 6.26, which shows the sketch of a

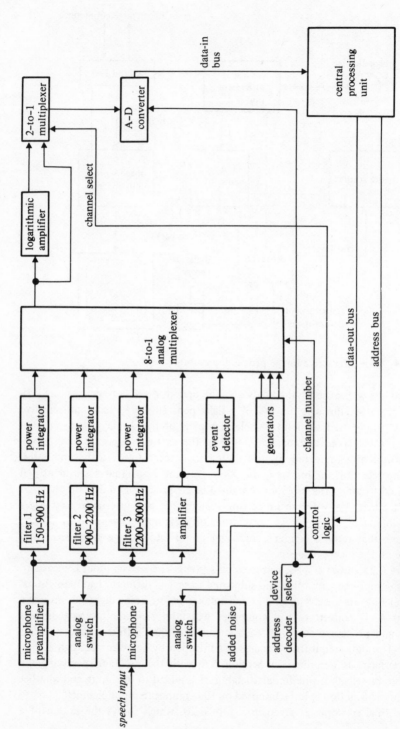

Fig. 6.25 A speech processing system

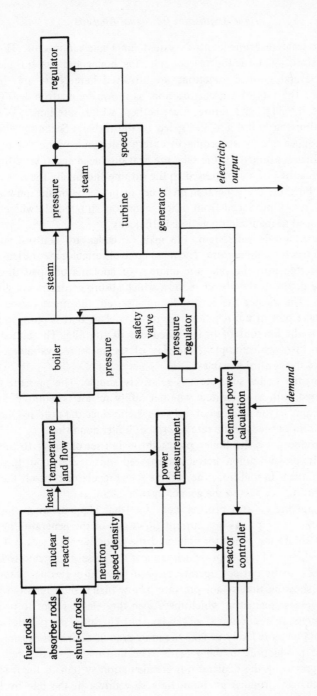

Fig. 6.26   Sketch of a nuclear generator control system

typical large-scale real-time control system for a nuclear reactor. Three main data sources are involved: the reactor pile, the boiler and the turbine-electric generator. Several control functions are involved here. First are the safety requirements. The rate of nuclear reaction, as shown by neutron density within the pile, and the pile temperature, must be kept within safe limits, as must the pressure within the boiler and the speed of the turbine. Secondly, the system needs to produce energy according to external demand, which determines the amount of fuel consumed by the pile, the heat produced and transmitted to the boiler, the quantity of steam passed to the turbine and finally the rate of power generation. Further, the generator must be 'in phase' with the power on the electricity network produced from other generators situated elsewhere. Finally, there is the need to maintain operational efficiency.

In the reactor pile subsystem, the main variables are neutron speed and density, and reactor temperature. The main control variables are the length of the fuel rods inserted into the pile, the amount of coolants pumped through the pile, and the depth of the absorber rods, which absorb neutrons and slow down the reaction. The deeper the fuel rods are inserted, the more fission material there is to take part in the reaction; the more coolant that is pumped through, the greater is the amount of heat carried out of the pile. The actual control process then involves a complex interplay of the measured variables, external demands, and the control variables.

Control in the boiler subsystem is relatively simple. The pressure measurements are used both to determine whether safety release is required, as well as to provide input to the reactor control subsystem, since the need for *future* heat production is influenced by the rate of *present* boiler condition.

Steam produced by the boiler passes through a set of valves to the turbine subsystem. Its pressure upon arrival is measured and used as input to the boiler subsystem control. In addition, the turbine speed regulator controls the amount of steam permitted to pass to the generator subsystem.

To illustrate the control process, consider what happens when the load on the output circuit drops. As less current flows out of the generator, less torque is needed to rotate the generator axle, and the rotor starts to accelerate. This is detected by the speed regulator, which reduces the passage of steam to bring the generator back to its correct operating speed (say 50 cycles per second). The information showing that steam pressure is now higher than required is fed to the boiler pressure regulator, which may open the release valves to produce an immediate pressure reduction if necessary, but (because safety release wastes heat) it would be more likely to take the more gradual course of causing a reduction of heat production in the reactor. The information that a reduction is desirable is passed by the demand power calculation system to the reactor controller, which can produce an immediate slow-down in the pile by inserting absorber rods deeper into the pile. However, if everything is operating smoothly a more gradual course of action will be taken. The rate of coolant pumping is reduced so that less heat is passed to the boilers, which will lead to a temporary heat build-up in the pile. However, at the same time the fuel rods are gradually withdrawn from the pile to reduce the amount of heat production, and perhaps the absorber rods are also adjusted to create a new optimum neutron density and speed balance. We see that an overall adjustment takes place over the whole control system, with step-by-step data being transferred continually and decision-making taking place in different subsystems. Information on all these control operations would be recorded continuously by the data logger associated with the reactor plant.

These are all complex examples of data acquisition systems and incorporate many of the components and techniques we have discussed in earlier chapters. The sequential actions of detection, measurement, processing, recording and control are seen to be carried out in many different situations. Other factors may be involved, such as the need for extreme safety which we saw in the last example and the requirement for back-up, but, in general, no system, no matter how complex, fails to break down into this elementary sequence of data acquisition operations.

References

1. DOEBELIN, E. O. *Measurement Systems: Application and Design*. McGraw-Hill, New York, 1967.
2. MILLMAN, J. and HALKIAS, C. S. *Integrated Electronics: Analog and Digital Circuits and Systems*. McGraw-Hill, New York, 1972.
3. BEAUCHAMP, K. G. and YUEN, C. K. *Digital Methods for Signal Analysis*. George Allen & Unwin, London, 1979.
4. MAGRAB, E. C. and BLOMQUIST, D. S. *The Measurement of Time-Varying Phenomena*. Wiley, New York, 1971.
5. LAWRENCE, D. E. and FENWICK, P. M. (eds.). *Data Acquisition and Real Time Systems*. University of Queensland Press, St. Lucia (QLD, Australia), 1971.
6. HANDSCHIN, E. (ed.). *Real-Time Control of Electric Power Systems*. Elsevier, Amsterdam, 1972.
7. HEALEY, M. *Principles of Automatic Control* (2nd edn). English Universities Press, London, 1975.
8. *Industrial Measurement Techniques for On-line Computers*. IEE Conference Publication No. 43, IEE, London, 1968.
9. *Organization and Management of Computer-Based Control and Automation Projects*. IEE Conference Publication No. 104, London, 1973.
10. *Trends in On-line Computer Control Systems*. IEE Conference Publication No. 127, London, 1975.
11. *Centralized Control Systems*. IEE Conference Publication No. 161, London, 1978.
12. LOCKS, M. O. *Reliability, Maintainability and Availability Assessment*, Hayden Book Company, Rochelle Park (NJ, USA), 1973.
13. TIETZE, U. and SCHENK, C. *Advanced Electronic Circuits*. Springer-Verlag, Berlin, 1978.
14. LOWE, E. I. and HIDDEN, A. E. *Computer Control in Process Industries*. Peter Peregrinus, London, 1971.

15. SMITH, C. L. *Digital Computer Process Control*. Educational Publishers, Scranton, Penn., 1972.

ADDITIONAL REFERENCES

BLASCHKE, W. S. and McGILL, J. *The Control of Industrial Processes by Digital Techniques*. Elsevier, Amsterdam, 1976.

JOHNSON, C. D. *Process Control Instrumentation Technology*. Wiley, New York, 1977.

SAYERS, B. McA, SWANSON, S. A. V. and WATSON, B. W. *Engineering in Medicine*. Oxford University Press, Oxford, 1975.

HILL, D. W. *Principles of Electronics in Medical Research*. Butterworth, London, 1973.

STACY, R. W. and WAXMAN, B. D. (eds.). *Computers in Biomedical Research* (Vol. 3). Academic Press, New York, 1969. (See also Vol. 4, 1974.)

CRUL, J. F. and PAYNE, J. P. *Patient Monitoring*. Excerpta Medica, Amsterdam, 1970.

HILL, D. W. and DOLAN, A. M. *Intensive Care Instrumentation*. Grune & Stratton, New York, 1976.

EWING, G. W. and ASHWORTH, H. A. 1974. *The Laboratory Recorder*. Plenum, New York, 1974.

CROMWELL, L., WEIBELL, F. J., PFEIFFER, E. A. and USSELMAN, L. B. *Biomedical Instrumentation and Measurements*. Prentice-Hall, Englewood Cliffs (NJ, USA), 1973.

Chapter 7

Microprocessors for Data Acquisition

7.1 Introduction to Microprocessor Hardware

Microprocessors are now an integral part of many engineering systems. They are of particular value to data acquisition and signal analysis systems because of their potential to satisfy many of the needs of such systems in data storage, processing and system control. In the present section, we shall introduce briefly the basic principles of microprocessor hardware. The subsequent sections will discuss the twin requirements of system interface and software development. Finally, several applications of microprocessors in data acquisition will be described.

We shall be considering, to a large extent, the application of a micro-processor as a microcomputer. Microcomputers differ from minicomputers and from even larger 'mainframe' computers, mainly because they are small and cheap and can be used for tasks where it would not be realistic or even possible to use the larger machine. Whilst considering the microprocessor for this purpose, it must also be realised that the microprocessor is simply a programmable component consisting of an aggregate of suitable logic elements, and whilst it can serve as a component in a microcomputer it is not itself a computer [1–5].

The microprocessor represents one of the more recent developments of integrated circuit technology. This technology now makes it possible to pack several thousand logic elements onto a small silicon wafer known as a **chip**. This process is known as large-scale integration (LSI). Analysis of the many different information processing systems has led to the partition of these logic circuits into self-contained modules which carry out well-designed functions. A micro-processor is an assembly of one or more LSI chips which form a **memory** used to store the data being processed as well as a set of **instructions** specifying *how* to process the data; and a **processing unit** which executes (i.e. carries out) these instructions. The various possible instructions a particular microprocessor is able to execute are called its **instruction set**, and vary from machine to machine, depending on the manufacturer, model, cost, complexity, etc. However, it is common to find instructions such as those needed for moving data from one part of the machine to another performing some arithmetic or logical operations on the data, or testing for some condition, e.g. whether a number is zero or non-zero, or whether number x exceeds number y, and then executing either of two alternative instructions depending on the result of the test. (This provides the microprocessor with decision-making capabilities.) A microprocessor also has I–O (input–output) **channels** which are used to bring data into the system from other devices or to send data out from the system. Instructions that do these are called I–O instructions. A **program** is nothing more than a collection of instructions which together achieve some particular purpose.

As we shall see, it takes some time for the processing unit to obtain data from the memory. Consequently, in order to speed up this operation for repetitive tasks it is desirable that the most frequently needed part of the data should be stored in some more directly accessible locations. This is why every processing unit will have a small number of high-speed **registers**. The processing unit is able to move data from register to register, or to perform manipulations of the data held in the registers, without having to invoke memory accesses. There are also instructions that move data between a register and the memory, or between a register and the I–O channels. Obviously, building more registers into a processing unit would tend to increase its processing capability. However, because of various constraints, microprocessors usually have no more than 8 or, in some cases, 16 registers.

These various elements form the essential components of a microprocessor. To see how they are interlinked and used, we will need to study their characteristics and operation in some detail. The following sections are arranged to provide this information. As an essential preliminary it may be useful if the reader revises, at this point, the background to the principles of computing and definitions of logic elements given earlier in Chapter 1.

7.1.1 THE MEMORY

At the present time, all microprocessors contain semi-conductor memories. The elementary storage element containing one binary bit is the **flip-flop**, which can be set to produce an output level having either a 0 (low voltage level) or a 1 (high voltage level) (see Section 1.7.1). Each flip-flop, therefore, contains one bit of information. Since it takes a number of bits to make up an instruction or a numerical value, the contents of a memory are grouped into a set of **memory locations**, each containing usually 8 or 16 bits. The locations are individually accessible. That is, one can read out the content of one location, or write new a content into a location, without affecting other locations in any way. An 8-bit location is said to contain a **byte** of information. The number of bits contained in each memory location is called the **word-length** of the machine. Most microprocessors have a word-length of 8 or 16, but longer word-lengths are beginning to be realised as technology develops further. However, the memory subsystem contains more than just a large number of flip-flop devices. It must also include, or be associated with, a certain number of control logic elements to enable the processor to read and write information correctly to the memory. In operation, the processing unit must be able to specify *which* memory location it wishes to access at a given time. This is achieved by assigning to each memory location an **address**, i.e. a serial number. The electronic components of the memory are organised in such a way that the memory is linked to a number of address and control lines (connected to its logic circuit), and to a number of data lines. To read the contents of a memory location, the processing unit must first place the address of that location on the address lines, and then send a pulse along the **read control line.** This pulse causes the data stored in the addressed location to appear on the data lines, and the processing unit can then connect the data lines to the desired destination for the data. To write into a memory location, the processing unit must first place the address on the address lines, the data on the data lines, and then send a pulse along the **write control line.** This causes the data to be connected to the particular memory location. The read and write operations are illustrated in Figure 7.1. The left-hand half of the diagram shows first, at the top, the position immediately before a read operation. An address 0110 is placed on the address line and a read pulse on the read line. After the read operation (bottom) the number stored in location 0110, namely 1011, appears on the data line. Before a write operation, shown at the top of the

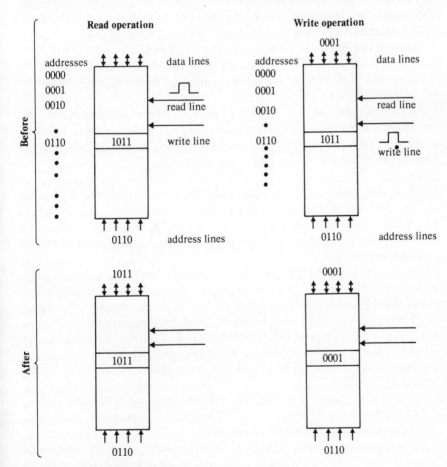

Fig. 7.1 Read and write operations

right-hand half of the diagram, the address 0110 is placed on the address line and the new number to be written, 0001, on the data line. After the write operation (bottom) the new number, 0001 replaces the earlier 1011 number in the addressed store location.

During the operation of the microprocessor, sequences of memory access operations, like those described above, occur automatically. When the processing unit is required to execute any instructions that involve reading from or writing to a particular memory location, the electronic circuits in the processing unit automatically generate the appropriate electric pulses that will activate the address, data and control lines. It is not necessary for the programmer to concern himself with such details apart from being aware of what is actually taking place.

To summarise the job of a memory within a microprocessor, we find that memories are required to perform a number of distinct functions:

1. Work space memory: This is required for the temporary storage of variables during a calculation and so need not be a permanent feature of the memory. It will require fast access and will be quite limited in size.
2. Data memory: This is used to store constants, tables of results, etc. This may be a temporary or semi-permanent requirement.
3. Program memory: This is used to store instructions which control the operation of the processor. Usually a permanent storage requirement.

Many different memory types, each known by its own special terminology, are available for use with a microprocessor to carry out these functions. It will be useful in the ensuing discussion of microprocessor operation, if we give here some simple definitions of these various types of memory.

1. Read-only memory (ROM): This is used for permanent storage of programs and data. It is not possible to change the data stored by writing new data to the memory. This is described as a **non-volatile** memory, since if the power is interrupted and then restored, the memory retains the information previously fed into it.
2. Programmable ROM (PROM): This is a special type of ROM which again is not alterable in normal operation by the microprocessor. However, it is possible to enter data into a new PROM by electrical means, but having done so once the information cannot subsequently be changed.
3. Erasable programmable ROM (EPROM): This is a ROM containing a special semi-conductor device which can be programmed to act as a conductor but which may be returned to its non-conducting state if required (see Section 7.3.4).
4. Electrically alterable PROM (EAROM): This is a more advanced form of EPROM and can be reprogrammed directly by the computer and the data is then stored indefinitely like the ROM.
5. Random-access memory (RAM) or read-write memory (RWM): This is used to store and read information in the way familiar with digital computers.

Data written into the memory will overwrite the previous contents. The information is lost when the power supply is interrupted and is thus described as a **volatile** memory.

7.1.2 THE PROGRAM COUNTER AND INSTRUCTION REGISTER

During the execution of a program, the processing unit has to fetch the instructions one by one from their locations within the memory. In order to carry this out, it is necessary to place the address of each instruction on the address lines and cause the memory to be read in order to obtain the instruction. Consequently, the processing unit must have some device to contain the address of the instruction that is *about* to be executed. This is provided by the **program counter**, which is an internal register. When one wishes the processor to run some program located in the memory, the starting address of the program must first be loaded into the program counter (PC) (this is usually carried out manually from switches located on the control panel). The start button is then depressed and the processing unit will obtain the first instruction from the memory and execute it. At the same time it will increase the value in the PC by 1 so that PC will now contain the address of the next instruction. After the execution of the present instruction is completed, the processing unit will use the content of the PC to find the address of the next instruction, and so on.

As mentioned earlier, it is possible for the processing unit to choose between two alternative instructions, depending on the result of some test we may build into the program. One choice is simply to access the instruction found in the next memory location by increasing the PC contents by 1. Alternatively, the instruction may be found elsewhere in the memory, and to execute this it is necessary first to load a new address into the PC. Thus, instructions (called **branches**) that choose between two alternative instructions simply decide whether or not a new value is to be sent to the PC. Figure 7.2 shows a typical program structure implemented with a branch. The condition $x = y$ is tested. If it is false the processing unit will go on to execute the following instruction. If it is true, the **branch instruction** will cause a block of instructions to be skipped and an alternative block, B, to be executed instead. However, regardless of which block is actually executed, eventually the processing unit will reach and execute instruction A. This structure is known as the IF . . . THEN . . . ELSE . . . construct and is a frequently employed programming tool.

After the processing unit fetches an instruction, it loads the instruction into a special **instruction register**. This register is connected to the **instruction decoder**, which analyses the instruction to see what operation is to be carried out, whether memory accesses are required, etc. Such information will then generate the necessary control pulses to put the instruction into effect. When the execution of one instruction is completed, other parts in the processing unit are activated and further control pulses are generated which cause the next instruction to be fetched, leading to yet another operation to be initiated.

We see that the two registers, PC and instruction register, relate to instructions

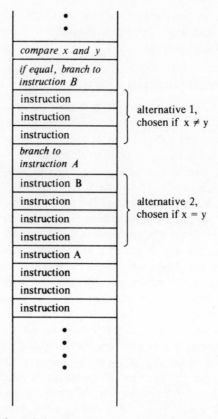

Fig. 7.2   The use of conditional and unconditional branches

in two different ways. One shows where to find each instruction, or *points to* each instruction; the other actually contains the instruction. The same can apply to data. We can have some registers actually containing data being processed, while other registers will contain addresses of data, and the processing unit will, when the need arises, fetch data from the memory via these registers. Registers that point to items of information, whether instructions, data or something else, are called **pointers**. When we fetch data from memory via registers, we are said to be using the registers in a **deferred** or **indirect mode** because they lead us to, but do not contain, data.

As described earlier, the basic function of the PC is to point to the instruction *about* to be executed. It does not point to the instruction being executed *now*, as the PC is always increased by one after an instruction has been fetched; once an instruction has been fetched, the memory location containing it is no longer required by the processing unit. This is why, at all times during program execution, the PC points to the instruction *below* the instruction currently being executed. This is an important fact, because it makes it possible to put the PC to another use.

Suppose we wish to use a numerical constant during the processing operations. The processing unit will need to store the constant in an easily accessible location. It is always possible to build a few registers into the processing unit specially for this purpose. However, the total number of constants we could store in this way would be quite limited, and it would be rather troublesome to have to keep a record of which constant would be needed by which instruction. It would be far easier if the constant could be placed next to the instruction which referred to it. Now, as the PC always points to the location below the instruction currently being executed, if we place the constant in that location, the processing unit will be able to fetch the constant also from the PC. In this way, our program will contain both instructions and relevant data. It is possible, in fact, to put several items of data required in sequence below each instruction. When the instruction first starts its execution, the PC will be pointing to the first item of data immediately below the instruction. After this has been fetched, the PC will be increased by 1 to point to the second item, and then the third item, etc. After all the data items needed by the present instruction have been fetched, the PC will be increased once again to point below the last data item, which is where we would put the next instruction, so that when execution of the present instruction finishes the processing unit will then be able to continue automatically with the next execution in sequence.

There are, of course, some restrictions on the kind of data items that we can put with each instruction; they must be known values at the time the program is written. Values which are to be created during the program execution cannot be stored and accessed easily in this way. Thus, constants are quite easily included in the instruction sequence. Also, we may not know the *value* of a particular data item required for an instruction, but we do know its address. It is then quite feasible to place the *address* below the instruction in the same way.

This is known as **absolute addressing** and is illustrated in Figure 7.3. The first stage (a) shows the position when the instruction is about to be executed. The processing unit will first use the PC to find the data address by placing the number 1101 found in the PC on to the address lines of the memory and will cause a read pulse to be generated. The second stage (b) shows this causing the memory location below the instruction, 0110, to be read, so that the address found, 0110, appears on the data lines of the memory. In the next stage, (c), the processing unit moves this number to the address lines of the memory and generates a second read pulse. In the final stage, (d), this is acted upon to cause the value of the data found, 1011, to appear on the data lines. This may appear rather complicated, but is typical of the way in which the programmer needs to consider programming a fundamental computing device such as a microprocessor. Fortunately, most modern microprocessors perform these operations automatically if the programmer uses the right type of instructions. Programming in what are known as the 'higher level' languages such as FORTRAN or PASCAL are not concerned with such basic operations, and will be considered later in this chapter. To recapitulate: the technique of placing constants below the instruction and fetching them during the execution of the instruction via the PC

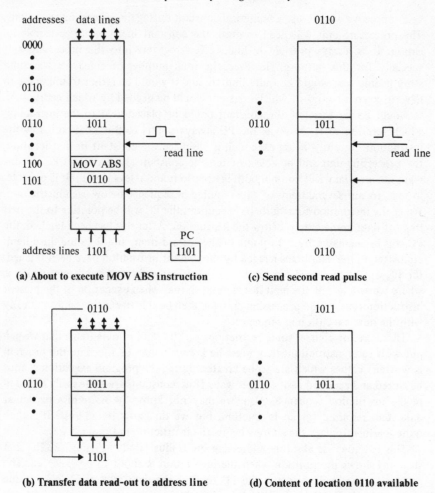

(a) **About to execute MOV ABS instruction**

(b) **Transfer data read-out to address line**

(c) **Send second read pulse**

(d) **Content of location 0110 available**

Fig. 7.3   Absolute addressing

is called **immediate addressing**. The technique of placing the address of some data item below the instruction and using this to fetch the data during execution is called **absolute addressing**, since the program writer is required to know the actual address of the data. As we shall see later, this requirement is difficult to satisfy when a program is complex and there are a great number of data items, but there are ways of getting around the problem. We also emphasise to the reader that, every time the processing unit uses the PC in order to fetch the next instruction or data item in sequence, it increases the value of the PC, so that it can use the PC later to find an item further down the stored information in the memory. This fact should be remembered as we shall need to refer to it later.

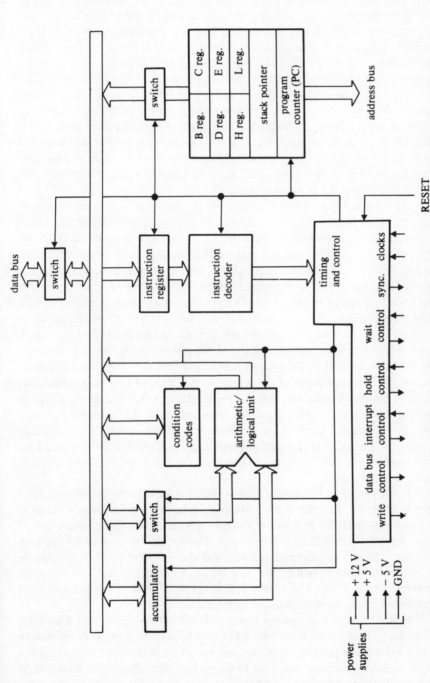

Fig. 7.4   Hardware configuration of Intel 8080 microprocessor (included by courtesy of Intel Corporation (UK) Ltd)

7.1.3   AN EXAMPLE

Figure 7.4 shows the hardware configuration of a popular microprocessor, the Intel 8080. The processing unit is packaged independently in a 40-pin (connectors) chip. Sixteen of the pins are to be connected to the address lines of the memory, while eight pins are for the data lines. The remaining pins contain such essential connectors required for power supply, ground, etc., as well as various control lines such as the memory write. Because of a shortage of pins, some necessary control pulses do not have their own connectors, e.g. the memory read pulse has to be generated by adding extra circuits! However, these detailed problems do not concern us here. With 16 address lines, an Intel 8080 processing unit is able to address up to 2^{16} individual memory locations. These are provided by joining one or more memory chips to the processing unit. We shall examine briefly the arrangements for interfacing later in this chapter. As may be inferred from the number of data lines, each memory location contains eight bits. We refer to the 8080 series as byte machines.

It can be noted from Figure 7.4 that the processing unit includes a program counter, an instruction register, a decoder, an arithmetic/logic unit which is used to perform data manipulations, and a number of registers. Some of the latter are required for special purposes, and may not be used by the programmer to store his data. With the Intel 8080, seven registers, identified as A, B, C, D, E, H and L, are available for access by the user. Instructions are provided which will move data from one register to another, from registers to the memory, or from memory to the registers. However, most arithmetic operations require the participation of the A register. There are, for example, instructions which will add a number taken from memory, or from another register to the number in A, but none that will add the same number to some other register; we shall see examples of this later. Because of the special status of the A register, it is referred to as the **accumulator**.

As the instructions are only eight bits in length, and each is required to specify *what* operations are to be carried out on *which* data, the instruction formats have to be designed carefully to accommodate all the necessary information. Because of this compact instruction format, any bits which are not required for one purpose are almost always re-allocated for other purposes, as we shall explain in due course.

The full list of instructions available with the Intel 8080 microprocessor is given in Table 7.1. In order to illustrate the function of some of these in transferring and manipulating data within the microprocessor, we will consider one of these, the MOV instruction, in some detail. This has the function of moving data between registers, or between registers and the memory. For convenience in register identification within the instruction format, the registers are given numbers, where B = 0, C = 1, D = 2, E = 3, H = 4 and L = 5; (A is identified separately as the special accumulator register).

The MOV instruction has the format 01DDDSSS, where D and S are each either zero or one. The three bits DDD may be seen to form a binary integer having a decimal equivalent between zero and seven. If the value is anything but six, it identifies that the register is to receive the data. Thus, if DDD = 000, it signifies that we wish to move a new value *into* register B. Similarly, the value

Table 7.1   **The Intel 8080 Instruction set: summary of processor instructions** (included by courtesy of Intel Corporation (UK) Ltd)

Mnemonic	Description	Instruction code								Clock Cycles
		D_7	D_6	D_5	D_4	D_3	D_2	D_1	D_0	
Move, load and store										
MOVr1r2	move register to register	0	1	D	D	D	S	S	S	5
MOV M r	move register to memory	0	1	1	1	0	S	S	S	7
MOV r M	move memory to register	0	1	D	D	D	1	1	0	7
MVI r	move immediate register	0	0	D	D	D	1	1	0	7
MVI M	move immediate memory	0	0	1	1	0	1	1	0	10
LXI B	load immediate register pair B & C	0	0	0	0	0	0	0	1	10
LXI D	load immediate register pair D & E	0	0	0	1	0	0	0	1	10
LXI H	load immediate register pair H & L	0	0	1	0	0	0	0	1	10
STAX B	store A indirect	0	0	0	0	0	0	1	0	7
STAX D	store A indirect	0	0	0	1	0	0	1	0	7
LDAX B	load A indirect	0	0	0	0	1	0	1	0	7
LDAX D	load A indirect	0	0	0	1	1	0	1	0	7
STA	store A direct	0	0	1	1	0	0	1	0	13
LDA	load A direct	0	0	1	1	1	0	1	0	13
SHLD	store H & L direct	0	0	1	0	0	0	1	0	16
LHLD	load H & L direct	0	0	1	0	1	0	1	0	16
XCHG	exchange D & E H & L registers	1	1	1	0	1	0	1	1	4
Stack ops										
PUSH B	push register Pair B & C on stack	1	1	0	0	0	1	0	1	11
PUSH D	push register Pair D & E on stack	1	1	0	1	0	1	0	1	11
PUSH H	push register Pair H & L on stack	1	1	1	0	0	1	0	1	11
PUSH PSW	push A and Flags on stack	1	1	1	1	0	1	0	1	11
POP B	pop register Pair B & C off stack	1	1	0	0	0	0	0	1	10
POP D	pop register Pair D & E off stack	1	1	0	1	0	0	0	1	10
POP H	pop register Pair H & L off stack	1	1	1	0	0	0	0	1	10
POP PSW	pop A and Flags off stack	1	1	1	1	0	0	0	1	10
XTHL	exchange top of stack H & L	1	1	1	0	0	0	1	1	18
SPHL	H & L to stack pointer	1	1	1	1	1	0	0	1	5
LXI SP	load immediate stack pointer	0	0	1	1	0	0	0	1	10
INX SP	increment stack pointer	0	0	1	1	0	0	1	1	5
DCX SP	decrement stack pointer	0	0	1	1	1	0	1	1	5
Jump										
MP	jump unconditional	1	1	0	0	0	0	1	1	10
JC	jump on carry	1	1	0	1	1	0	1	0	10

Mnemonic	Description	Instruction code								Clock Cycles
		D_7	D_6	D_5	D_4	D_3	D_2	D_1	D_0	
JNC	jump on no carry	1	1	0	1	0	0	1	0	10
JZ	jump on zero	1	1	0	0	1	0	1	0	10
JNZ	jump on no zero	1	1	0	0	0	0	1	0	10
JP	jump on positive	1	1	1	1	0	0	1	0	10
JM	jump on minus	1	1	1	1	1	0	1	0	10
JPE	jump on parity even	1	1	1	0	1	0	1	0	10
JPO	jump on parity odd	1	1	1	0	0	0	1	0	10
PCHL	H & L to program counter	1	1	1	0	1	0	0	1	5

Call

Mnemonic	Description	D_7	D_6	D_5	D_4	D_3	D_2	D_1	D_0	Clock Cycles
CALL	call unconditional	1	1	0	0	1	1	0	1	17
CC	call on carry	1	1	0	1	1	1	0	0	11/17
CNC	call on no carry	1	1	0	1	0	1	0	0	11/17
CZ	call on zero	1	1	0	0	1	1	0	0	11/17
CNZ	call on no zero	1	1	0	0	0	1	0	0	11/17
CP	call on positive	1	1	1	1	0	1	0	0	11/17
CM	call on minus	1	1	1	1	1	1	0	0	11/17
CPE	call on parity even	1	1	1	0	1	1	0	0	11/17
CPO	call on parity odd	1	1	1	0	0	1	0	0	11/17

Return

Mnemonic	Description	D_7	D_6	D_5	D_4	D_3	D_2	D_1	D_0	Clock Cycles
RET	return	1	1	0	0	1	0	0	1	10
RC	return on carry	1	1	0	1	1	0	0	0	5/11
RNC	return on no carry	1	1	0	1	0	0	0	0	5/11
RZ	return on zero	1	1	0	0	1	0	0	0	5/11
RNZ	return on no zero	1	1	0	0	0	0	0	0	5/11
RP	return on positive	1	1	1	1	0	0	0	0	5/11
RM	return on minus	1	1	1	1	1	0	0	0	5/11
RPE	return on parity even	1	1	1	0	1	0	0	0	5/11
RPO	return on parity odd	1	1	1	0	0	0	0	0	5/11

Restart

Mnemonic	Description	D_7	D_6	D_5	D_4	D_3	D_2	D_1	D_0	Clock Cycles
RST	restart	1	1	A	A	A	1	1	1	11

Increment and decrement

Mnemonic	Description	D_7	D_6	D_5	D_4	D_3	D_2	D_1	D_0	Clock Cycles
INR	increment register	0	0	D	D	D	1	0	0	5
DCR	decrement register	0	0	D	D	D	1	0	1	5
INR M	increment memory	0	0	1	1	0	1	0	0	10
DCR M	decrement memory	0	0	1	1	0	1	0	1	10
INX B	increment B & C registers	0	0	0	0	0	0	1	1	5
INX D	increment D & E registers	0	0	0	1	0	0	1	1	5
INX H	increment H & L registers	0	0	1	0	0	0	1	1	5
DCX B	decrement B & C	0	0	0	0	1	0	1	1	5
DCX D	decrement D & E	0	0	0	1	1	0	1	1	5
DCX H	decrement H & L	0	0	1	0	1	0	1	1	5

Add

Mnemonic	Description	D_7	D_6	D_5	D_4	D_3	D_2	D_1	D_0	Clock Cycles
Add r	add register to A	1	0	0	0	0	S	S	S	4

| | | Instruction code | | | | | | | | Clock |
Mnemonic	Description	D_7	D_6	D_5	D_4	D_3	D_2	D_1	D_0	Cycles
ADC r	add register to A with carry	1	0	0	0	1	S	S	S	4
ADD M	add memory to A	1	0	0	0	0	1	1	0	7
ADC M	add memory to A with carry	1	0	0	0	1	1	1	0	7
ADI	add immediate to A	1	1	0	0	0	1	1	0	7
ACI	add immediate to A with carry	1	1	0	0	1	1	1	0	7
DAD B	add B & C to H & L	0	0	0	0	1	0	0	1	10
DAD D	add D & E to H & L	0	0	0	1	1	0	0	1	10
DAD H	add H & L to H & L	0	0	1	0	1	0	0	1	10
DAD SP	add stack pointer to H & L	0	0	1	1	1	0	0	1	10

SSS identifies the register *from* which the data item comes, again provided the value chosen is not six. Thus, the instruction 01011100 will move the content of register H (= 4) into register E (= 3). The previous content of E will be lost by this operation while that of H is not altered. (In this sense, a MOV instruction can be said to *copy* information from one register to another.) The register identified by SSS is the *source* of data, while that identified by DDD is the *destination* for the transferred data. The two bits 01 constitute the **operation code** or **opcode**. In this particular case, 01 identifies the instruction as a MOV instruction. The other three possible opcodes, 00, 10 and 11, have more complicated meanings. They have to be looked at in conjunction with other bits in the instruction word in order to find what they mean. We shall see a small number of examples later.

What if either DDD or SSS = 110? These numbers are reserved to indicate *register to memory* moves and not the *register to register* moves that we discussed earlier. If DDD = 6, the content of the source register is to be copied into a memory location; if SSS = 6, the information stored in a memory location is to be read and loaded into the destination register. (Note that DDD and SSS may not both be 110, otherwise we no longer have a MOV instruction.) But the processing unit must provide the address of the memory location involved. The assumption here is that the address is already in the H and L registers, which together contain 16 bits, just enough for an address. The H register contains the *high-order*, or more significant half of the address, and L the *low-order* half. The two registers together form a memory pointer for data, just as PC is a pointer for instructions.

But how do we ensure that the H and L registers contain the correct address? We can move the two halves of the address into these registers by means of the MVI, or **move immediate** instruction, which has the format 00DDD110, which specifies that the byte immediately below the instruction is to be moved to the destination. We see that SSS = 110, indicating that the source is in the memory rather than a register. However, because this instruction has a different opcode from the MOV instruction, the processing unit obtains the memory address from the PC rather than the H and L registers, ensuring that the source information will be fetched from below the instruction being obeyed.

To take an example, suppose we wish to move the content of location 1025 into the accumulator. We must first load the address into the H and L registers. The high half of the address is 00000100 while the low half is 00000001. Thus, the following set of instructions accomplish the operation:

00100110	(move the byte below into register no. 4)
00000100	(high half of address)
00101110	(move the byte below into register no. 5)
00000001	(low half of address)
01111110	(move byte whose address is in H and L into register no. 7)

Quite obviously, programming in this manner would become much too complex to be carried out accurately by the programmer, except for very simple tasks. This is why one requires some software development aids in order to be able to write realistic programs for microprocessors. However, before we can discuss these it is necessary first to consider several other hardware aspects of the microprocessor.

7.1.4 BRANCHES AND CONDITION CODES

We mentioned earlier the branch (or jump) instructions. It was said then that these instructions permit the selective execution of parts of the program depending on conditions created by previous instructions. Suppose we wish to execute one out of two program segments depending on whether x exceeds y. Then we would first employ a CMP (compare) instruction, and follow this with a **jump-if-greater** instruction. While the actual details vary from machine to machine, and also depend on the task in hand and on the programming style of the person, the above represents a fairly standard way of carrying out the operation among microprocessor designs.

It should be clear, however, that the processing unit must have some mechanism to record the results of the **compare** instruction in order that the branch instruction following it can use the information to decide whether or not a jump is to occur. We refer to 'results' here because several pieces of information have to be recorded, such as whether $x - y$ is equal to 0, whether it is greater than 0, whether it is greater than 2^8 (a **byte overflow**, since this last condition means that $x - y$ is too large to be stored in one byte), whether it has even parity, and perhaps some other condition the programmer wishes to record. By recording a number of such information items, the mechanism allows us to follow the compare instruction with a variety of different branch instructions, depending on the programming requirements, e.g. jump-if-equal, jump-if-greater, jump-if-overflow, etc. The device used for this purpose is called the **condition codes register**, which contains a number of bytes, each indicating by their value (1 indicating yes or set and 0 indicating no or clear) that the previous instruction created some particular condition. The condition codes register of Intel 8080, for example, is able to record five conditions: ZERO (i.e. the previous instruction produced a zero result), SIGN (a negative result in 2's complement arithmetic), CARRY, PARITY and HALF CARRY. Note that the register may be affected by more than just the compare instructions. For example, if we add two numbers or perform some logical operations on them, information about the result is immediately recorded in the condition codes register. Also, if we move a number from one location to another, or indeed do anything that affects any data anywhere, information about the number is recorded. Thus, if we load a new value into the accumulator, and it happens to be zero, then the ZERO bit of the condition codes register will be set, as well as the PARITY bit (since zero has even parity), whereas the SIGN and the two CARRY bits will be clear, since the number is not negative and the process of moving it did not produce any carry.

By making use of this additional information, we can thus use branch instructions in conjunction with a variety of other instructions to achieve quite complex programming operations. The full advantages of this condition codes register will be appreciated only after the programmer has had some experience in programming a microprocessor in this fundamental way, although, of course, as suggested earlier, there are software development aids which use these and other mechanisms without the programmer even being aware of their operation [6, 7].

7.1.5 SUBROUTINE JUMPS AND THE STACK

During the operation of a program one often finds that the same set of instructions is required in several places. For example, the generation of the sine or cosine of some angle would be a frequent requirement in many signal processing applications. Similarly, the need to examine the status of an input device and to move any data available into a given register also occurs frequently. If we were simply to write the same set of instructions again and again, we would be wasting much program storage space. Consequently, it is usual to place such a program segment once only in a fixed location in the memory. When the sequential execution of instructions in the program reaches the place where the segment is required, the processing unit performs a jump to the separately stored segment, executes it and then returns to the previous program at the instruction below the jump. Such separately stored program segments are called **subroutines**, and a jump to a subroutine is usually termed a **subroutine call**. The program making the call is called the **main program**. The subroutine is made to return to the main program upon completion of its set of instructions. Figure 7.5 illustrates the call/return process.

Subroutine calls and returns have to be handled differently from branches. This is because at the time one writes the subroutine, one does not know to which address the return is to be made, as the subroutine may be called from several different places in the main program. Thus, at the time that the main program calls the subroutine it also has to store the address of the next instruction

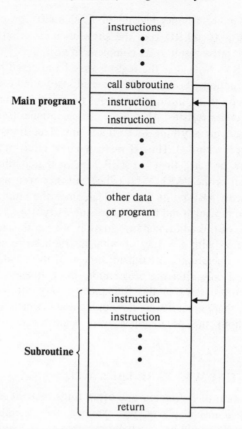

Fig. 7.5   Subroutine call and return process

in the main program at some place accessible to the subroutine. This operation is usually called **saving** the return address.

Consider, however, the case when the subroutine itself calls another subroutine. When the second subroutine is being executed, it is necessary to have two saved return addresses, one to return to the first subroutine, and the other to return, later, to the main program. In general, it is possible to have a long chain of subroutine calls, and the number of addresses that need to be saved increases as one goes further along the chain. The situation is illustrated in Figure 7.6, where we have program A calling subroutine B, and then B calling C and C calling D. During the execution of D, three return addresses need to be preserved, from the time of each subroutine call. Note that the last address saved is the first to be used (to return from D to C), and the first address saved the last to be used (when returning from B to A). Such a structure, in which the most recent arrival will make the first departure (or last-in-first-out), occurs frequently in computing, and is normally provided for in the processor hardware design in the form of a **stack**.

A stack consists of a number of registers or equivalent register locations within

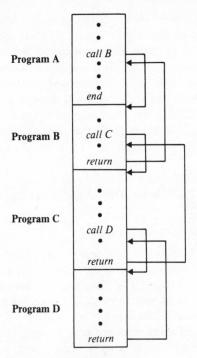

Fig. 7.6   Chained subroutine calls and returns

the memory. A series of registers would be limited in number, and in many micro-processors, including the 8080, the stack is simply a block of memory locations specially designed to contain data and, as such, are virtually unlimited in number. There is a special register called the **stack pointer** which always contains the address of the most recent entry in the stack. When one makes a subroutine jump, the address of the instruction below the subroutine jump is transferred and stored in the stack. When the subroutine finishes, the return instruction causes the processing unit to retrieve the return address from the stack and the stack pointer is altered to point to the next most recent arrival. Again these operations are automatically performed by the processing unit. All that the pro-grammer has to do is to insert subroutine jump instructions at the right places in his main program, and return-from-subroutine instructions at the ends of the subroutines.

Before leaving the subject, let us look a little closer at what we mean by 'return address'. As we know, during the execution of each instruction, the PC points to the instruction below it. Normally, when the execution of one instruction is completed, the processing unit will simply fetch the next instruction via the PC. However, a branch instruction will cause a new address to be loaded into the PC, so that the next instruction executed comes from elsewhere in the program. In a subroutine jump, the same thing happens. The starting address of the subroutine has to be loaded into the PC in order that the instructions in the subroutine can be executed. However, the previous content of the PC is not simply erased, since later, when the subroutine finished, the instruction sequence must return to the main program and the instruction below the subroutine jump instruction will need to be executed. This is why the subroutine jump instruction has to cause the current value of the PC to be saved in the stack. Later, when the subroutine finishes, the return instruction will simply retrieve this value and put it back into the PC, causing the processing unit to obtain its next instruction from the main program again. In short, 'return addresses' are simply previous contents of the PC, saved in the stack at the time of the subroutine calls.

7.2   Input—Output and Interfacing

7.2.1   INPUT—OUTPUT TRANSFERS

On first inspection I—O instructions appear to be just data-move instructions of a special kind. Some machines (such as the Intel 8080) have special I—O instructions that will cause the transfer of data along a specified I—O channel, e.g. IN 300 will cause a byte of data to be moved from the device interface at the end of the channel number 300 into the accumulator register. Other machines do not even need special transfer instructions. Instead, I—O devices are assigned special memory addresses so that a MOV instruction involving these addresses will actually cause I—O transfers directly from device interface to the memory. (Such mechanisms for I—O handling are said to be **memory mapped**.) However, in actual fact, the situation is much more complicated. When the processing unit executes instructions involving the memory it exercises full control, since the response time of memory to read or write control signals is well defined and the data can be transferred with no delays of any kind. I—O devices, on the other hand, take much longer, and usually uncertain or variable amounts of time to respond. Some devices are controlled by external agents, such as a person pushing buttons at his terminal device. This is why I—O control signals are much more complex. Also, the processing unit must have some means of observing the status of each I—O device. By **status**, it is meant whether the device is in a position to accept or transmit data, whether it has as yet assembled a complete byte or word of information for transmission, or completed the reception of a byte, etc. — in other words, its 'state'of operation. Depending on its status, different control signals are needed and only certain I—O transfers are possible at the time. This is why I—O handling must involve both hardware and software designs: the hardware necessary to communicate the control and status information between the processing unit and the I—O devices, and sometimes, to generate or convert such signals; and the software controlling the processing unit in order to recognise these signals and make decisions based upon them, and to transmit further control information in turn.

The questions might be asked, 'Why have the microprocessor designers failed to include in their instruction sets special instructions that could perform status testing or control signal generation of various kinds? Why leave so much to the users?' In fact, it would not have been particularly difficult to implement such instructions. The main obstacle is the lack of external connections (pins) from the microprocessor to connect such status and control pulses with external devices. Even the largest microprocessing units in use today have no more than 48 pins. Thus, even though the processing unit may be capable of generating elaborate control pulses internally, there are no means of connecting such pulses to the I–O devices! Instead, the processor unit has to issue its control signals in 'shorthand' form, which are later expanded to provide detailed pulses using the hardware contained in the I–O system. Fortunately, most of such signal conversion tasks are well defined so that manufacturers can design a number of standard components for carrying these out. As a consequence, it is only necessary to interface the appropriate standard components between the I–O devices and the processing unit to enable most of the more usual I–O operations to be controlled through the microprocessor instructions.

In this brief introduction we shall not be able to study either the hardware or software aspects of microprocessor I–O mechanisms in any detail. Instead, we describe the basic techniques and component types in common use and refer the reader to other books and manuals for additional information, some of which are listed at the end of this chapter [8, 9].

7.2.2 BUFFERS

Let us start with a simple case, an output device that accepts data one byte at a time. This could, for example, be a low-speed printer which, for each different byte received, advances its carriage one slot position and prints some symbol specified by the value of the received byte. (Obviously there must be some system for identifying each symbol with a particular value. The most commonly adopted system for microprocessors is the 8-bit ASCII code [11].) However, even for such a simple case as this, certain interfacing requirements need to be met. Firstly, we find that the output device will take some time to respond and initiate the output operation. Therefore, there must be some storage mechanism to retain the output data which is accessible to both the processing unit and the output device. The mechanism used is another register, the **output buffer register**. The processing unit loads each byte in turn into this register so that the output device can access the data from there. Secondly, the output device takes a certain time to complete the output operation, during which period the processing unit must refrain from transferring the next output character to the device, as this would interfere with one currently being output. This is achieved by providing yet another register, a **status register**, which is again accessible to both sides. This register is in some sense similar to the condition codes register, except that its contents reflect the condition of the *output* device: for example, whether the device is busy or free. Before the processing unit transfers a byte into the buffer register of the device, it must first test the bit of the device status register

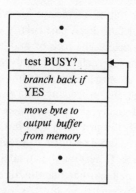

Fig. 7.7   Idle loop for output transfers

designated for this purpose. This is known as the BUSY bit. If the bit is zero, then the processing unit initiates the output. As soon as the output device control mechanism receives the byte, it changes the status to BUSY, making the BUSY bit a logical one. The next transfer will be delayed until the first transfer is completed, when the BUSY bit will be made zero by the circuits contained in the output device. Figure 7.7 illustrates this system, which is called the **idle loop** method for output transfers. The processing unit executes an instruction repeatedly to test the BUSY signal in the output device; if it is non-zero, the device is not available to perform the next operation so that, instead of carrying out any output transfers, the program merely jumps back to the previous test instruction to try the BUSY bit again and again as long as the device is still busy. If, however, the bit is zero, then the next instruction, which is an output transfer, will be executed. The term 'idle loop' arises from the fact that the processing unit spends its time repeatedly testing for the BUSY bit, waiting for the device to become free, instead of using the period to do something more productive.

Input handling is more complex because even a simple input device has at least three different states. It may be idle, busy reading in data, or having just finished reading in data. The last state is not the same as idle, since if the input device is idle then it will have no input data ready for the processor, whereas in the last state it will have. The idle condition permits an input operation to be initiated. However, some time is needed before the device succeeds in obtaining the data. So the BUSY state has two different stages: BUSY but data not READY, and BUSY and data READY. Thus, the status register of an input device has to have at least two condition bits: BUSY and READY. Moreover, there also needs to be a control bit accessible to the processing unit to indicate that the input operation can commence. (In the output device, moving a byte into the buffer register of the device would automatically initiate an operation. Input operations have to be started specifically by the processor.) The set of instructions needed for the complete input process is shown in Figure 7.8. Here the program first ensures that the device is free. It then orders a control pulse to be sent to the START READ bit of the input device. This immediately turns on the BUSY bit

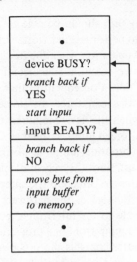

Fig. 7.8   Program for simple input control

and causes the device to accept data from its input devices, i.e. card reader, papertape reader or a terminal, by placing the data read in into the buffer register from which it may be accessed by the processing unit. Concurrently with these operations, the program tests the READY bit repeatedly to see if data are now available. When the READY bit turns to one, the program proceeds to the next instruction, which transfers a byte from the device buffer register to the accumulator, whereupon further processing can take place. It is possible with some devices to transfer whole blocks of data at a time, instead of single bytes. This requires larger buffer registers. For output, the processing unit would move the data block into the output buffer, together with a number of specifying the size of the block. For input, the processing unit would initiate a block read instruction and wait for the data to be READY. The input device would move the data available into its buffer and indicate the number of bytes contained in the data to the processing unit, which would then execute data move instructions to transfer the data into the memory. We shall see later, however, that block transfers are usually handled in a different way in order to prevent the processing unit initiating repeated MOV instructions, which is a very time-consuming process.

7.2.3   INTERRUPTS

The use of a programmed idle loop for I–O transfers has two disadvantages. First, no productive work is performed while the processing unit waits for the I–O device to become READY. Second, it is not possible to handle multiple devices using this method, particularly if some of the devices are controlled externally, since all I–O transfers have to be initiated by the program. In most realistic systems, it should be possible for I–O devices to initiate I–O transfers at the time the devices (and not the processing unit) require them. Further it is desirable that the processing unit should be able to *start* an I–O operation and then proceed immediately to perform other work while the I–O device is BUSY, returning to finish the I–O operation when the device becomes READY. Such a system would thus handle two activities concurrently, one related and the other unrelated to the I–O operation. Consequently, there needs to be some mechanism by which the I–O device can force the processing unit to suspend any activity unrelated to I–O when the device requires it and to cause a jump to the I–O related activity. Such a forced jump is called an **interrupt**.

Interrupts differ from subroutine jumps in several ways. Subroutine jumps are controlled by the program, since they occur at specific places in the program as specified by the programmer. In contrast, interrupts are forced upon the processing unit by the fixed design of the hardware components, and occur at times and places unpredictable to the interrupted program. They are, however, similar, in that both involve a jump from one program segment in the memory to some other segment. The latter controls the processing unit for some time but eventually finishes, whereupon a jump back to the original program segment occurs. Both have to be provided for in the design of the processing unit, the former by the provision of the subroutine jump and subroutine return instructions, and the latter by inclusion of a number of control lines. One of these lines is defined as an **interrupt request line**, along which an I–O device requiring to cause a jump to a program related to the device can send an interrupt signal. A second line is designated as an **interrupt acknowledge line**, along which the processing unit informs the device that the interrupt requested is allowed to take effect. Other signals may need to be passed, since the I–O device may need to notify the processing unit of the address of the I–O related program so that the processing unit can cause a jump to that location. (In the Intel 8080, the I–O device is also required to provide an *instruction* to the processor, usually a special jump instruction called 'restart'; we omit other details.)

In a practical system, we may have simultaneous interrupt requests from several different devices. Most manufacturers provide a component called the **priority interrupt unit**, which permits differing priorities to be assigned to all the devices in use. This operates by suppressing all but the one with the highest priority from the simultaneous requests received and services this highest priority interrupt first. Some of the components will also save the received but unsatisfied requests until the earlier interrupt has been serviced by the processing unit, at which time other interrupts are permitted to take effect. Further components may be added to permit the processing unit to specify a **threshold priority**. For example, the processor may place a number into a special register; to interrupt the processing unit, the I–O device must have a priority higher than that number. This prevents unimportant I–O requirements from disrupting essential program executions. Like the subroutine jump instruction, an interrupt causes the present values of the PC to be saved in the stack so that, at a later time, the execution of instructions can resume by restoring the former value of the PC. However, other information also needs to be saved. Since an interrupt could occur between the execution of an arithmetic or logical instruction and a branch instruction, the contents of the condition codes register have to be saved so that, when the processing unit returns from an interrupt to the branch instruction, the correct information will be available in the condition codes register to allow the processing unit to decide whether a branch operation is needed. Some microprocessors cause the contents of *all* the registers to be saved in the stack, since they form part of the data of the interrupted program. When the execution of the interrupt-related program is completed, a return-from-interrupt instruction causes the restoration of former values in all the registers, including the condition codes and the PC, so that the interrupted program is able to continue from its point of suspension with all its previous data intact.

There are, in fact, a variety of different interrupts. Besides I—O devices, error detection units can also cause interrupts, so that the programs generating the errors may be suspended to permit a given error-handling program to be executed. Such interrupts usually have the highest priority. Finally, we have **software interrupts** which are really a special kind of subroutine call; again we omit details.

7.2.4   DIRECT MEMORY ACCESS

Earlier we mentioned block transfer devices. Usually such devices are connected to the microprocessor system through **direct memory access** (DMA) mechanisms, which transfer bytes between the I—O device and selected locations in the memory automatically, without requiring continuous control and supervision by the processing unit. Such devices have their own address registers. When the processing unit wishes to initiate a block transfer, it places the address of the first memory location that contains (or is to receive) the data into the address register of the I—O device. For output devices, the number of bytes to be transferred is also given to the device. The processing unit sends the START I—O signal, and causes a jump to some other program not directly related to the I—O activity. The I—O device then proceeds to perform the transfer under its own control. When the device requires either to fetch a byte from the memory or to store a byte into the memory, it will place the content of its address register onto the address lines of the processing unit. Control signals are generated that will 'lock out' the processor so that it cannot temporarily gain access to the memory. Further control signals cause a transfer to occur between the device and the memory. The address register of the device is increased by one and the number of bytes to be transferred is modified to prepare for the next byte transfer. All this happens during a short period of time. The processing unit plays no part in this DMA transfer, except for being temporarily prevented from accessing the memory during the transfer period. When the transfer of a whole block is completed, the I—O device interrupts the processing unit, which is then able to use the data for processing or to start yet another block transfer at another memory address location.

DMA mechanisms are expensive and complex. They are therefore employed only for high-speed devices such as magnetic disc units that transfer fairly large amounts of data during a given period. In such cases, the cost is justified because the method frees the processing unit from the detailed, byte by byte control of I–O transfers. Its intervention is limited to starting the block transfer and to processing a complete block contained in the memory.

7.2.5   BUSES AND OTHER COMPONENTS

We described in a previous section how the processing unit requires a set of data lines, a set of address lines and a number of control lines. The data lines require connection to all the devices and all the memory modules. Furthermore, the data lines need to transmit pulses accurately in both directions, regardless of the distance between the transmitter and the receiver or how many other devices there are in between. This is why it is necessary to have specific electronic systems to control the lines to ensure accurate communication, free of degradation, interference, secondary pulse reflections, etc. Such a system is called a **bi-directional bus**. A microprocessor system may contain separate buses for data, address and control and for communication between processing unit and the memory, or between the memory and the I–O devices.

Other components frequently employed in microprocessor interfacing are **address decoders** and **multiplexers**. Address decoders receive as an input a memory address. A number of output lines are provided connected to various I–O devices. When an address is received, the decoder will send out one control pulse to the I–O device to which that address refers. The decoder thus carries out a device-selection function. The multiplexer, on the other hand, provides centralised traffic for a number of devices, such that data received from them will enter the microprocessor system along a common path, reducing the number of paths required.

7.3   Software Development for Microprocessors

7.3.1   ASSEMBLY PROGRAMMING

We saw earlier that instructions are strings of binary numbers representing opcodes, register numbers, addresses, etc. It would be extremely difficult to write programs directly in this binary form. The variety of operations, and therefore opcodes, is usually large and it would be necessary for the programmer to refer constantly to his manual in order to determine the opcode of every operation. Further, for any task of moderate complexity a large number of data items would be required. While it is possible for the programmer to undertake the laborious task of assigning memory locations to data, and then to write the program referring to the addresses for all the data, the work would be extremely slow and prone to error. Finally, if new data or new instructions are added to an existing program, or to a program in the process of testing and **debugging** (error correction), then the addresses of most data items and instructions will be subject to change, so that the whole program will virtually have to be rewritten with old addresses being replaced by new ones.

This is why it is more usual to write programs in an indirect fashion. Instead of identifying each data item with a memory location, the programmer gives it a symbolic name such as X1, X2, NUMBER, VECTOR, STRING, etc. This provides several advantages. The names are more meaningful and easier to remember than actual memory addresses. Also, if for any reason data are moved to new locations, the names do not change and instructions employing those names need not be altered. Similarly, each instruction can be given a symbolic name (often called the **symbolic address**) so that branches and subroutine jumps can specify the destination of each transfer symbolically. Yet another improvement is to give every operation a symbolic name (which is called a **mnemonic**). It might also be recalled that in the Intel 8080 microprocessor even the registers have been assigned symbolic names (A, B, C, D, E, H and L).

A program of this type is called an **assembly program**, which has to be translated into the corresponding **machine code** comprising binary strings which can be loaded into the memory and executed. The assembly program is known as a **source code** since it originates from the programmer; the machine code is referred to as the **object code**, since it represents the final desired product. The conversion is carried out by an **assembler**, which is itself a computer program. This is designed to accept the source code provided by the programmer and provide an output in the form of the equivalent object code, usually on some output device such as a magnetic disc, magnetic tape or paper tape, for subsequent input into the microprocessor through one of its input devices.

However, before the translation can take place, it is first necessary for the assembler to allocate memory locations to the programmer's data. This obviously means that the programmer must have some method of specifying what data he has available and must be able to define the memory requirement for each data item. An alphanumeric character, for example, can be stored in one byte, a small integer in two bytes, a low-precision real number or large integer in four bytes, and a vector in a large number of consecutive memory locations. The notations employed for such purposes are called **pseudo codes**, to distinguish them from normal instructions. The assembler then allocates the required memory locations to all the data, and computes the address corresponding to each symbolic name. It is then able to convert all assembly program instructions, along with all their symbolic addresses, into corresponding binary strings. To take an example, the Intel 8080 assembly languages permits the following instructions for data movements:

MOV r1, r2	moves the content of one register (r1) to another (r2).
LDA XYZ	moves the content of a memory location having the symbolic name XYZ to register A (accumulator).
STA XYZ	performs the reverse of above.
MVI r, 1234	loads the constant 1234 into register r (this is a move immediate, the constant being

	located immediately below the instruction, as explained earlier).
LXI B (or D or H), 1234	loads a 16-bit constant into a pair of registers, either B and C, or D and E, or H and L.
LHLD XYZ	moves the content of memory location XYZ to the register H and that of next memory location to register L.
LDAX B (or D)	loads the content of a memory location pointed to by the 16 bits in registers B and C or D and E into the accumulator.
MOV r, M	loads the content of a memory location pointed to by the 16 bits in registers H and L into the accumulator.
STAX B (or D) and MOV M, r	are the reverse of the above two.

If any of the above instructions are used in a program, the computer will convert them into the appropriate binary strings (object code), and will also attach to the instructions any addresses they need for locating data.

Before leaving the subject of assembler programming it is worth mentioning the problems of compatibility between assembler instruction sets written for a particular microprocessor and another in the same manufacturers' range or even from another manufacturer. Within the same range, it is now becoming common for software to be what is known as 'upwards compatible' for a range of microprocessors of similar design. Thus the instruction code for an 8-bit microprocessor will be found to form a subset of the larger instruction set available with a 16-bit microprocessor, which in turn forms a subset for the 32-bit microprocessor instruction set. Thus, a program developed for an 8-bit machine may also be run on either the 16- or 32-bit machine with no further development. To a lesser extent, this compatibility is also found between different manufacturers' equipment so that some measure of software interchangeability is possible. The true 'common ground' between microprocessors of different manufacturers is tending towards the use of high-level languages, which we consider in Section 7.3.3.

7.3.2 SELF- AND CROSS-ASSEMBLERS

We said earlier that an assembler is required to translate the source programs into object code. Being a computer program itself, an assembler has to be executed on some machine. If this is the same microprocessor which is to receive the object code, then we have a **self-assembler**. However, most microprocessors for which we wish to develop software are not capable of running the assembler. The main problems are associated with the I–O facilities and memory requirements. An assembler usually requires a considerable amount of memory in order to function efficiently, and the microprocessor systems do not normally have adequate storage. However, since memory is relatively cheap, this problem would not itself prevent assembly in most cases. A more serious difficulty is the requirement of reading in the source program to the memory and then producing the object program in some machine-readable form which can later be entered again into the machine for execution. Further, since programs require a great deal of modification and testing before they can be used, it is necessary to have some facility for program changing, test execution and observation of the results by the user. Unfortunately, microprocessor systems seldom have adequate I–O capabilities, for the very reason that microprocessors are so cheap. It makes little economic sense to connect thousands of dollars of terminal, printer, papertape and disc units to a ten dollar processing chip, especially as once the program has been thoroughly tested, many of these I–O devices will no longer be required, since the functions in which we are interested will involve microprocessor control of equipment for data acquisition rather than interaction with human programmers.

Owing to this I–O limitation, software is rarely developed on the same microprocessor that will later use it. Two operational methods are commonly used. One is to have a number of identical microprocessor systems but equip only one with sufficient I–O equipment and memory for software development. Programs are written, stored, translated and exhaustively tested on that machine and then transferred to the other microprocessors for productive work. Usually, the program transferred is in pre-loaded ROM modules. That is, the development machine produces the object code and stores it, and then reads it back as an input into a machine incorporating the ROM. The ROM memory chips may be removed from the machine, still retaining the stored program thus produced, taken to the other microprocessors and connected to the memory bus. Occasionally, systems are encountered where the development machine produces the output as an object program on papertape. Later the production machines may read the tape into their own memories and execute the code. The latter system obviously would require an extra tape reader for the use of the production machines, as well as a pre-loaded tape-read program (a **loader**) in the machines for loading the object code, but these would be justified if the production machines needed to run a variety of different programs.

A different scheme is to run a **cross-assembler** on a different computer. This may be a large-scale system used as a general purpose computer, or yet another, larger microprocessor. Whichever system is used, it will need to have good input–output facilities, especially for on-line program storage, editing and debugging and good error diagnostics to help the programmer. It would be possible to write the cross-assembler in a high-level language and this is very often done. In contrast, a self-assembler on a microprocessor has to be, at least in early stages, written in machine code. However, cross-assemblers are not without their disadvantages. First, it is still necessary to be able to load object code into the microprocessor, and thus an input device is still needed. Second, there is the problem of program testing. When the development machine is identical with the production machines, software fully tested on one machine will be capable of execution immediately on the others. When a cross-assembler is employed, program testing is made considerably more difficult, especially since in such cases the microprocessor will have little I–O facilities, so that it is not easy to observe the effects of the program during execution. If the object code is to be stored later in ROMs then the process becomes even more complicated. One must first produce the object code on the large machine, and input the code via an output media into the microprocessor which contains a writable memory for testing purposes. Finally, the corrected object code is used to control a microprocessor containing a ROM to yield the final written memory which then can be transferred physically to the production machine.

It may be said that software development for microprocessors is nearly always a time-consuming, expensive and also frustrating exercise. The cost of software development, including the hardware used and programming effort employed, far outweighs the cost of the production hardware. Thus, generally it is necessary to be able to spread the cost of programming over a large number of units. The consolation is that, once the software is working, then the incremental cost of additional units becomes extremely low, as one needs to purchase little more than a few LSI chips and to connect them together.

7.3.3   HIGH-LEVEL LANGUAGES AND PACKAGES

The difficulties of programming in the microprocessors' own language (machine code) have been mentioned earlier. As with the evolution of the larger computing equipment, the first technique to be developed and implemented to overcome some of these problems was the assembler language. With this, the instructions for the processor are given mnemonic names, and a special program, called the assembler program, is used to convert these into sets of machine-readable code that the processor can act upon. However, assembler code programming demands a certain level of skill which many users of the microprocessor are unwilling to acquire for very good reasons – the principal one being the effort involved in what is to the user (who is generally not a professional programmer) an unproductive exercise. Consequently, the use of a **higher-level language** for the preparation of microprocessor programs is now highly developed.

High-level languages allow the actual algorithms needed to solve a problem to be written directly. Unlike machine code and assembly code, it is not necessary to refer to the detailed structure of the machine such as size or number of registers, since such languages rely instead on abstract considerations such as variables, expressions and procedures. This gives two important advantages to the programmer. In the first place, the programmer need not remember the precise arbitrary details of the machine for which he is writing and can direct his full attention to the operations he is trying to program. Secondly, the algorithms he uses in a high-level language are not related to any particular design of machine and can therefore run on any computer or microcomputer for which a suitable compiler exists. (A **compiler** is a computer program which converts the high-level language statements into a series of machine-code instructions and, of course, is written for a particular machine.) Such programs are called **portable** and have obvious advantages over programs written for one machine only. The most important high-level languages for use with the microcomputer are PASCAL,

BASIC, and FORTRAN, all of which enable the user to develop his software in very much the same way as would be carried out for a large-scale computer. Discussion of the features and use of compilers would take us outside the scope of this book and the interested reader is referred to the references given at the end of this chapter [7, 9–11].

As noted earlier, additional facilities are necessary to support high-level language compilers, both in terms of hardware and in manufacturer supplied software. The use of a cross-assembler represents one way of preparing the microprocessor object code from the high-level program. However, the continued improvement in memory and back-up storage facilities, such as the floppy disc and cassette, enables many microprocessors to contain a resident high-level compiler, even though this may lack some of the facilities of the version used in the larger machines. Thus the tendency is towards programming techniques which render the task of program preparation very little different from that for the full-sized computer.

Although the microprocessor has no need of an operating system such as found in the general purpose computer, with its special language for entering and controlling jobs to be run, it is often provided with a restricted suite of control software. This may include a file management routine that maintains a record of the locations of programs stored in its associated tape or disc peripherals and an editor which enables programs to be entered and modified interactively from the keyboard. Other specialised package software is available to carry out a wide variety of tasks, including many of the signal processing and data processing operations which previously required either specially designed and interconnected hardware or the use of a much larger general purpose computer.

7.3.4 DEVELOPMENT SYSTEMS

The design and construction of a microprocessor-based system is concluded by the debugging, testing and commissioning of the hardware. With microprocessors, a number of techniques have been evolved to enable both the hardware and software to be successfully 'debugged' together. These techniques are known as development systems and a number of basic approaches are used.

One of these is the **simulator**, which is a computer program written to run on a large computer and which allows the programmer to check his program without having access to the microprocessor at all. This program takes all the instruction code of the particular microprocessor as its input data and simulates the complete operation of the microprocessor, including its attachment to the various peripherals and interface connections that would be used in the complete system. This method gives excellent fault-finding information to enable errors in the microprocessor program to be corrected. It also provides a completed and corrected program in machine code for entering into the PROM of the microprocessor, together with information on the program execution time when run on the microprocessor itself and other details. A major difficulty with this technique is the slow execution of the code under simulation conditions which makes the testing unrealistic where real-time control of the microprocessor I—O is important.

An alternative method is to use a second microprocessor as a monitor processor to provide information on the running of the user's program in his microprocessor. The microprocessor on which the final program will run is known as the target microprocessor and is a minimal system consisting only of those essential components necessary for the application. A larger host microprocessor or minicomputer is used to store the program as files on magnetic disc. Operations are carried out on this stored information, using an assembler program and an **editor program**. The editor is used to manipulate these files easily when an error is discovered. This is carried out as an interactive process so that the user can monitor the changes he makes on the screen of a visual display unit. A number of operations or **edit commands** can be caused to act on a single line of instruction, and positioning commands can enable this line to be moved about within the program text and displayed on the screen. A more sophisticated approach involves editing by context, i.e. by searching the text and causing the program to stop when the line containing the word defined is found within the program. In either case, the line can be inspected and corrected by means of suitable edit commands.

Larger computers, such as mainframes or minicomputers, are provided with banks of indicator lights and switches which are used to set up addresses to indicate the state of registers and generally to permit manual control of programming operations. The microprocessor and most microcomputers do not have these facilities and instead the functions are taken over by a software monitor program. This is a small program supplied by the manufacturers as a ROM permanently contained in the memory of the microprocessor. Whenever the microprocessor is switched on or reset, the monitor program is entered automatically at its starting point and the user can communicate with it via a terminal. The facilities provided by the monitor vary, but most monitors incorporate the features listed below.

1. Memory examine and deposit: this allows a portion of the memory content to be examined and changed if necessary.
2. Load binary: this allows binary code generated by the assembler program on the host computer to be loaded into the microprocessor memory ready for execution.
3. Enter program: this enables the loaded program to be executed.
4. Break points: here the user sets the number of breakpoints within the program where normal sequential execution would be suspended. When one of these breakpoints is reached in the program, then the user can examine the state of the registers, etc., to check their contents before permitting execution to be recommenced.

All these testing tools are software aids to microprocessor development. The designer will also need to have some control over the hardware operation in a

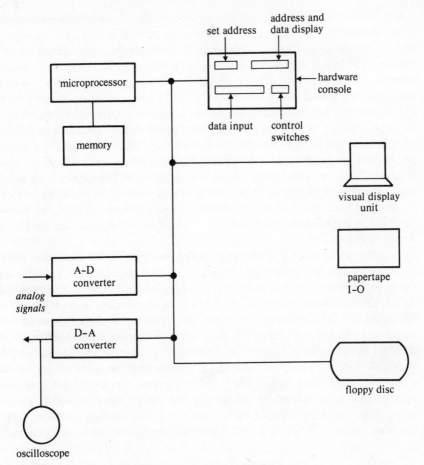

Fig. 7.9   Hardware console microprocessor configuration

more direct way and this is provided by the **hardware console**. Owing to the cost involved, only a limited range of fault-finding is attempted but this is more direct and immediately apparent to the operator than would be software simulation and monitoring. A typical configuration is shown in Figure 7.9.

These generalised front panel systems are designed with features which can be exploited during initial system development, including debugging and fault-finding in operational systems which lack panel controls and display mechanisms. The kind of facilities available are indicated in Figure 7.9. The switch controls on the panel can be used to load addresses, transfer data into the memory, to examine a memory location and read the register contents. The instructions in the program can be obeyed slowly in a step-by-step sequence while the address or register contents are examined and the program started or stopped as required. The state of operation can also be indicated by appropriate indicator lights, e.g. run, wait, transfer, etc.

An EPROM programmer may also be connected, as well as interface circuits for connection to an even larger development system. It will be recalled that one of the possible memory systems is the erasable programmable read-only memory (EPROM) which is contained in most microprocessor development systems. This cannot be reprogrammed directly by the system in which it is used and requires a special unit known as an **EPROM programmer** to carry this out. This contains a semi-conductor device which permits a series of electric charges to be transferred to the memory storage locations corresponding to the bit pattern to be stored and then effectively isolates the charges from leakage due to any further electrical input. Removal of these charges prior to reprogramming is brought about by exposing the charge-storage device to ultra-violet light which causes the charges to migrate back into the silicon substrate of the device.

An invaluable piece of test equipment developed especially for the microprocessor is the **logic state analyser**. The general arrangement for the analyser is shown in Figure 7.10.

A number of input probes are provided to enter signals from the microprocessor equipment under test. These are usually connected to the bits contained in the address bus and to the control bus signals. In some analysers, the data bus contents are entered as well. The signals are matched with switches contained on the front panel so that some combinations of them can be used to enable the memory to record all the inputs. When the input conditions are detected as being identical with the switch settings (0 or 1), then the memory is activated. Thereafter, sets of input values are stored in the memory, one set for each controlling clock pulse. When the memory is completely full then subsequent data are admitted by overwriting the earliest recorded memory word input, with the latest received value. One use for the analyser is as a 'digital historian' so that, for example, microprocessor operation can be arrested when

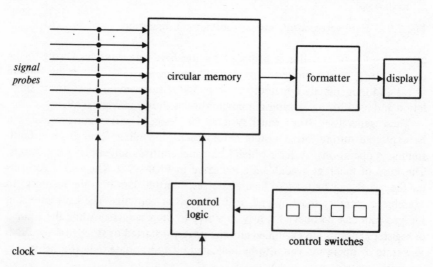

Fig. 7.10   A logic state analyser: general arrangement

a particular instruction address is reached and a number of the preceding instructions are displayed on the display screen.

Associated with the display unit is a **display formatter**. This is designed to reconstruct timing diagrams or present digital displays. These latter may present a choice between displaying lists of information in different codes, such as hexadecimal for address information or binary information on the control lines.

A more sophisticated form of display is the **memory map**. This shows in one overall display the general operation of a program. The contents of the address bus are used as control input. Two D–A converters are contained in the formatter with their outputs coupled to the x and y driving circuits of the display unit. The most significant bits of the monitored data are applied to one of the D–A converters and the least significant bits to the other converter. As the addresses are accessed the CRT beam moves and its instantaneous position gives an indication of the location in the memory of the particular instruction being operated.

We conclude this section with brief reference to the recent development of microprocessor system design kits for user applications. These are essentially single board microcomputer systems, including CPU, some memory, I–O facilities external equipment interfaces and software. A small keyboard and LED display contained as an integral part of the board enables direct insertion, examination and execution of the user's program. The design kits bring together in restricted form many of the development tools which we have been discussing and which are valuable for the construction of a single and dedicated application device such as a signal processor or an equipment controller. The board contains space for insertion of additional memory and interface components, A–D and D–A converters and other logic elements to suit a particular application.

Software often derived from a fairly extensive manufacturer's library can be selected and included with the kit in the form of pre-programmed ROM modules which are simply plugged in to the appropriate area of the board. This is a new approach to software acquisition which will expand and take precedence over secondary storage entry systems, such as cassette tapes and discs as the capabilities of bulk semi-conductor memories are increased.

7.4 Microprocessors in Data Acquisiton Systems

7.4.1 INPUT–OUTPUT INTERFACING

By far the most important applications of microprocessors in data acquisition are as intelligent interfaces between different I–O devices and between I–O devices and other computers. Some examples were given in earlier chapters. As the reader will now realise, I–O handling is a complex process, requiring frequent status testing, simple decision-making, and transfers between various concurrent activities, whether subroutine calls or interrupts. On the other hand, extended chains of arithmetic or logical data manipulations are seldom required. This is why it is efficient to design one's system such that I–O activities are handled by microprocessors, which are well equipped to perform such conceptually simple, but physically complex, tasks. Many of the modern computer peripherals, such as high-quality terminals, graphics displays, remote batch stations, terminal concentrators, graph plotters, etc., will include microprocessor controllers, in order to simplify the part the large central processor needs to play in I–O processing.

The use of microprocessors in I–O control ranges from the very simple to the fairly elaborate. As an example of the former, let us consider microprocessor control of a LSI ten-bit A–D converter designed on a single chip. The converter would be marketed as a standard component so that no alteration would be possible. Besides the ten output lines for data, the chip would also have various control input lines, such as clock, start, conversion, hold data, etc., and status lines such as busy (see Fig. 7.11). Assuming that the microprocessor is a byte machine and has an eight-bit data bus, it is first necessary to design some switching circuits which will selectively connect either the top two bits (with six leading zeros) or the less significant eight bits to the data bus. The various control and status lines also have to be connected appropriately. In particular, the timing pulses provided by the microprocessor-controlled clock determine the rate at which conversions will proceed, and the processing unit can, in fact, disable the A–D converter by stopping the clock connection. For each conversion, the processor has to send a control pulse along the start conversion line, and wait for the busy signal to be turned off, before initiating two transfers to bring the two parts of the output into the processing unit.

The above description applies to a system where the microprocessor devotes itself to the control of the A–D converter, which is likely to be a relatively low-rate one. For more elaborate systems operating at higher speed, either interrupt

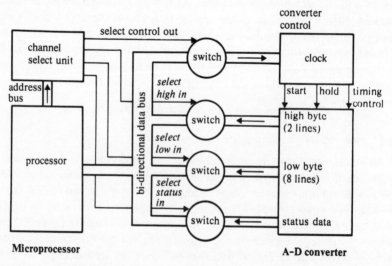

Fig. 7.11　Interfacing a ten-bit A–D converter to an eight-bit machine

or DMA mechanisms will be added, especially if there are a number of high-speed A–D converters, which will probably be multiplexed and share the same DMA interface for data input.

Another simple application of microprocessors is in data display systems. The processing unit will handle one or more A–D converters as described above, as well as data display outputs such as numerical displays, oscilloscopes or a combination of the two. The system may also include a simple or elaborate control panel, or perhaps a keyboard for the operator, which are both input devices and need program control. Besides I–O handling, the microprocessor will also perform the functions of data conversion and buffering. As mentioned in an earlier chapter, binary-to-decimal conversion is an extremely common requirement in computer system design. So is integer-to-floating-point conversion. Some simple arithmetic might also need to be carried out in order to combine the input of several channels and to produce other meaningful results. In buffering, we store previous values acquired for subsequent display, possibly combining earlier and later values to produce new results.

A further application lies in data storage systems. With an intelligent microprocessor in control, it is possible to connect a variety of data storage devices, possibly analog but more likely digital, to the system. The microprocessor can be arranged to accept data from the input devices and convert them to the required format for data storage (possibly rearranging the sequence of the data and performing some error checking, and inserting additional information for identification purposes) before finally writing them out onto the storage media. A more complex microprocessor-controlled storage device was outlined in Chapter 4.

Last but not least, microprocessors provide an excellent means of interfacing data acquisition devices to a large computer. Besides taking over the detailed I–O control operations, microprocessors, with their low-cost memory modules, provide extensive buffering capabilities. The measured data can be stored and arranged in the microprocessor system, and sent to the central computer only when large and complete blocks of data have been accumulated, and then only when the central computer is ready to process them. In this way, the computer's data storage and data manipulation facilities are not unduly taxed, and more efficient operation of the CPU can be obtained.

7.4.2 PRE-PROCESSING

The need for pre-processing and the various operations carried out on data before they can be processed were discussed in Chapter 3. There we considered mainly analog operations, but of course it is also possible to pre-process data digitally and the microprocessor forms a valuable method of carrying out this operation. For example, calibration may be performed easily by means of a stored table. The correct value corresponding to each raw input value is computed and stored, usually in a read-only memory chip. During the data acquisition process, the microprocessor is programmed to extract the calibrated value corresponding to a given input value. Since the input value is generally contained within the range 0 to $2^n - 1$ for some n (depending on the word-length of the A–D converter), the position of the calibrated value for input x is computed by adding x to the starting address of the table. On many microprocessors there is an addressing mode that makes explicit address computation unnecessary; this is the **index** or **register displacement** mode. Here the value of x, the table index (since it specifies which entry of the table is needed), is placed in a register, while the starting address of the table, known to the assembler, is placed behind the instruction. The processing unit is able to check that the instruction is in index mode, and will automatically fetch the table starting address and add it to the content of the register to find the memory address.

Trend removal is another pre-processing task readily implemented on a microprocessor, since the computation steps are relatively simple. By computing the mean and mean-square of a set of values and inserting these into a simple formula, a linear expression is produced that approximates to the input. Subtracting the value of this from each input yields its deviation from the 'overall behaviour' or trend. The trend values and the deviations provide separate information about the data. Digital filtering can also be implemented on a microprocessor. As shown in Chapter 3, filtering may be either recursive or non-recursive. Both operations require multiplication of several pre-computed coefficients into successive elements of the input or previous output and summing of the products. While such operations can be specially implemented, using a set of shift registers together with multipliers and adders, implementation on a microprocessor offers greater flexibility, with a greater choice of filters and coefficients. The coefficients are usually pre-computed on a large general-purpose computer and later fed into the microprocessor. When the coefficients have been decided once and for all, they can, of course, be stored in read-only memory.

We can also use the microprocessor for the digital reconstitution of analog signals. Input signals are digitised and manipulated in some way, and then converted back into analog form by D–A converters controlled by the microprocessor. An important application of this method is in time-scale changes; the microprocessor can store the digital values and convert them at a different rate from the input rate. This allows one to stretch or compress a function along the time axis without changing the shape of the function, a result difficult to achieve with fixed processing electronics. Simply storing the function and reconstituting it later may be a useful exercise, since achieving this by analog means may require circulating delay lines, whose accuracy and consistency leave much to be desired, or the more expensive alternative of the variable-speed analog magnetic tape recorder described earlier.

7.4.3 INTEGRATED SYSTEM CONTROL

As discussed in the previous chapter, an integrated data system consists of data measurement facilities, a control unit to process the input, and output facilities that operate on the results to perform functions on the data source. Conceptually, therefore, there is little difference between such a system and a microprocessor equipped with appropriate I–O devices, and a considerable amount has been written about such systems [12–16]. The main limitation on the viability of a microprocessor-controlled system lies in the relatively low data-manipulation capability of microprocessors. Since an integrated data system generally has a number of I–O devices that directly take part in the control function, as well as additional units, such as data display consoles, operator keyboard, data logging (recording) equipment for record keeping and error recovery, and possibly facilities for the storage and loading of software, the processor controlling the devices will usually have little spare capacity for data processing purposes. Consequently, most microprocessors will not be able to produce control signals at the speed required, except for relatively simple systems, i.e. a system with only a small number of devices, or one whose control function requires little data manipulation. However, the situation is changing. The development of integrated circuits in the last decade has produced a period of very rapid change, which will continue into the future. Only ten years ago, integrated circuits in production had an active element density up to 100, but by 1970–1 the single chip-calculator with a density of 1000 became available. Today, with the advent of the 64K random access memory (RAM) the active element density per chip can be in excess of 100 000. Densities of a million or more elements per chip are predicted and this will lead to memories in the multi-million byte region to be produced. This development is also affecting the word-length for the microprocessor. Sixteen-bit word-length units are now available and 32-bit microprocessors are possible. Longer word-lengths mean that larger, more accurate integers and real numbers may be processed quickly, whereas in earlier processors high-precision computation would automatically demand multiple instructions to combine multi-byte values. Also, the longer word-length enables more information to be transferred between different parts of the machine, especially between I–O devices and the processing unit, using the same number of instructions. With increased memory size, large programs and more data may be stored. Not only does this enable more complex processing procedures to be employed, but processing of the data can be carried out faster, by making use of large tables of pre-computed results located in the additional storage space and employing algorithms to reduce processing time. A further microprocessor development is found in the content of the instruction sets which are being expanded to include new, more complex data-manipulation operations. All these developments mean that microprocessors are becoming more powerful, taking over the role previously held by minicomputers, so that in the future they may well be able to satisfy many of our needs for integrated data systems control.

Together with this improved hardware, manufacturers are also investing more resources in providing sophisticated software, both to facilitate software development by users and for special applications. As mentioned earlier, microprocessors with simple operating systems, program editors and high-level language compilers are now beginning to be available at low cost. Special-purpose programs for fast Fourier transform and digital filtering are also developed for some systems, as well as complete microprocessor data logging or process control systems, marketed with all necessary hardware and software.

Because of the high cost of developing new microprocessor systems, semiconductor manufacturers are increasingly considering the advantages of copying existing complete computer systems. These could be minicomputers or mainframes, and the process would carry the obvious advantages of making use of existing support software. At present this is not easy to realise owing to the complexity of even a minicomputer system and the quite different organisational systems found there. The increasing density of elements on a single chip and large-scale memory availability does not, however, preclude this sort of simulation being attempted in the 1980s.

7.4.4   CHOICE OF A MICROPROCESSOR

Because of the great variety of available processors, the choice of a system specially suited to one's requirements is a daunting task. However, it is not our intention to comment on the merits or otherwise of specific models. Instead, we offer the following general considerations for the reader's guidance.

Package versus in-house development
Users wishing to have a system that has been well proven in actual application and can be made operational for his use in minimal time will naturally wish to have a complete, ready-made package. However, unless his application is very similar or identical to that of a large number of other customers, it is unlikely that he will find a package exactly suited to his needs. In-house development of particular systems, on the other hand, is both complex and time-consuming, and is not advisable unless the user is well equipped with both human and material resources, in laboratory facilities, computer availability (for cross-assembly or cross-compiling) and in testing equipment. Further, development of special systems will not be cost-effective unless the intention is to duplicate the resulting system many times in order to distribute the initial cost. For average users the most advisable approach would be to obtain a relatively general-purpose package that could be adapted to suit special purposes with relatively little effort. Special attention should be paid to software development tools included with the package.

Centralised versus distributed systems
When designing a system for controlling a number of concurrent activities, one can either divide these among various interconnected processors, or centralise them in a more powerful processor; both options cause complications. Whether one should adopt the first or second approach depends on the amount of interaction occurring between the activities. If the second approach is adopted, it is necessary to use a microprocessor system having a good real-time operating system for multi-programmed software development. For a distributed system

it is desirable to adopt a master–slave configuration, where one processing unit is responsible for allocating all system resources, including memory space and I–O devices, for the use of other 'slave' processors. It is the function of the master to fill the memory modules with the correct instructions and data, and to activate a slave processor for each program, with appropriate resources. In both cases, special attention should be paid to (a) communication between activities and protection against subsystem malfunctions, which should not be allowed to bring the whole system down (i.e. render it completely inoperative); similarly, care should be taken to prevent subsystems from accessing program or data they are not entitled to interfere with.

Special-purpose hardware versus software

The economics of digital electronics at the present time means that purchasing special-purpose hardware is more cost-effective than the user developing specialised software, since the hardware vendor is able to spread his development cost over a large number of units, whereas the incremental cost of each unit is small. In-house hardware development, on the other hand, is favourable only when there are special factors, such as an algorithm that is easy to implement in parallel hardware but difficult to carry out in serial instruction execution. In all cases, however, one must consider whether the hardware employed will increase or decrease the data-handling requirements. Hardware that produces an output greater than its input imposes additional requirements on system data transfers, and hence on the interface hardware and I–O handling software. This factor could often mean that the special-purpose hardware causes more activity than it saves. Also, whereas software can be readily modified, hardware is difficult to alter.

General considerations

In choosing a manufacturer, one should consider the capability of each machine for data manipulation and storage, the range of I–O devices, interrupt handling capabilities in relation to the number and speed of I–O devices, manufacturer-provided system and application software, and maintenance. In choosing a particular system configuration, one should consider the needs of the system in both production mode and in software development mode. The more software needs to be developed, the more sophisticated will be the I–O and program storage requirements. Finally, the need for support facilities for both hardware and software work cannot be stressed too highly. These include good documentation, adequate information on the product and project changes and adequate maintenance aids or service.

References

1. HILBURN, J. L. and JULICH, P. M. *Microcomputers/microprocessors: Hardware, Software and Applications*. Prentice-Hall, Englewood Cliffs (NJ, USA), 1976.

2. BARNA, A. and PORAT, D. F. *Introduction to Microcomputers and Microprocessors*. Wiley, New York, 1976.

3. SOUCEK, B. *Microprocessors and Microcomputers*. Wiley, New York, 1976.

4. PEATMAN, J. B. *Microcomputer-based Design*. McGraw-Hill, New York, 1977.

5. McGLYNN, D. R. *Microprocessors – Technology Architecture and Application*. Wiley, New York, 1976.

6. AGRAWALA, A. K. and RAUSCHER, W. G. *Foundations of Microprogramming Architectures, Software and Applications*. Academic Press, New York, 1976.

7. LEAHY, W. F. *Microprocessor Architecture and Programming*. Wiley, New York, 1977.

8. LESEA, A. and ZAKS, R. *Microprocessor Interfacing Techniques*. Sybex, Berkeley (Ca., USA), 1976.

9. LEVENTHAT, L. A. *Introduction to Microprocessors: Software, Hardware, Programming*. Prentice-Hall, Englewood Cliffs (NJ, USA), 1978.

10. LEWIN, D. *Theory and Design of Digital Computers*. Nelson, London, 1972.

11. ASPINALL, D. *The Microprocessor and its Applications*. Cambridge University Press, Cambridge, 1978.

12. RAO, G. V. *Microprocessors and Microcomputer Systems*. Van Nostrand Reinhold, New York, 1976.

13. BIBBERO, R. J. *Microprocessors in Instruments and Control*. Wiley, New York, 1977.

14. *IERE Conference on Microprocessors in Automation and Control*. University of Kent, England, September 1978.

15. GARLAND, H. *Introduction to Microprocessor System Design*. McGraw-Hill, New York, 1979.

16. KLINGMAN, E. E. *Microprocessor System Design*. Prentice-Hall, Englewood Cliffs (NJ, USA), 1977.

ADDITIONAL REFERENCES

17. ZAKS, R. *Microprocessors, from Chips to Systems*. Sybex, Berkeley (Ca., USA), 1977.

18. OSBORNE, A. *An Introduction to Microcomputers*. Osborne Books, Berkeley (Ca., USA), 1976.

19. ASPINALL, D. and DAGLESS, E. L. *Introduction to Microprocessors*. Pitman, London, 1976.

20. LIPPIATT, A. G. *The Architecture of Small Computer Systems*. Prentice-Hall, London, 1978.

Chapter 8

Remote Data Acquisition

8.1 Introduction

In the present chapter we provide a survey of data acquisition methods carried out over distance. Such remote data acqusition processes are implemented to handle various special circumstances. Some examples are given below.

1. The data sources may be beyond our reach, such as astronomical objects, space vehicles, or the interior of a live animal or human body. In military operations, the targets to be observed may be in areas under the control of hostile forces. Obviously, the only way of acquiring data from such sources is to perform the observations from a distant location.
2. The data sources may be reached separately, but are distributed over scattered locations, whilst analysis must be performed on the total sources of the data and the information transmitted to a central location for storage and analysis. One such example is the acquisition of meteorological information.
3. There may be only one data source. However, this may be situated in an inconvenient location, such as under water, over a mountain top, or in a remote jungle, and the cost of installing, maintaining and operating the complete equipment at that site will be prohibitively high. In such cases, it is more cost-effective to place only part of the data system at the remote location and to establish a data link between this and the rest of the system which would be situated in a more convenient location.
4. In many agricultural or geographical studies the data sources are numerous and distributed over a large area, e.g. vegetation covering part of a continent, and it is necessary to obtain an overall picture of the collective behaviour of the data sources, without being concerned about the individual data sources. By situating the measuring apparatus some distance from the area and having a sufficiently wide field of data acquisition, a collective view of the composite data source can be obtained.
5. The data sources are mobile, e.g. wild animals. It is then necessary to attach some measuring devices to the source or sources and arrange for the data acquired in this way to be transmitted from the sources to a central observation point. A similar requirement is found for field equipment carried in the back of a truck or in a doctor's briefcase, where occasional measurements from scattered data sources have to be transmitted to a central system for

real-time analysis or recording. An example applicable to our second situation would be that of immobile patients being treated in their homes.

6. There are also cases where measuring instruments may be moved away from the location of the data recording–analysis system to some place where less external interference is found, e.g. astronomical observations from a satellite where it is found necessary to reduce signal attenuation and interference caused by passage of the signal through the Earth's atmosphere.

We are said to be performing **remote sensing** if the measuring devices are situated away from the data source, and **telemetry** if the measuring devices are situated away from the data acquisition system. There may be systems that include both. Remote sensing may be either **passive** or **active**. In the former case, the measuring devices merely receive signals produced by the data sources; in the latter, the system sends out its own signals and measures the signals returned from the objects under observation. Radar and sonar, using radio and sound waves, respectively, are well-known examples in this second category.

When determining which particular remote data acquisition process is best suited to the requirements of a problem, the following considerations must be taken into account.

1. The choice of the **information carrier**: In passive remote sensing one usually has a choice of signals to be chosen for observation, since data sources often emit a variety of signals, such as visible light (which may be of various frequencies), infra-red rays, ultra-violet rays, microwaves and other kinds of radiation. One must therefore decide which frequencies are the most useful for extracting information about the data sources. For example, visible light is useful for distinguishing between different types of vegetation or soil observed from the study of satellite photographs, infra-red rays for analysing temperature variations, X-rays and gamma rays for detecting the presence of radioactive minerals, whereas exploration of metallic ores may use both radio waves, to detect changes in conductivity in the Earth's surface, and light, because of colour differences between the ore and its surrounding material. In short, the choice of an information carrier is heavily influenced by the nature of the data sources and the particular kind of information sought. The second factor here is the capacity of alternative carriers to penetrate the medium existing between the data sources and the measuring devices or data system. For example, radio waves can travel through clouds and blizzards, making them suitable for all-weather observations from the air or on the ground. On the other hand, they cannot travel far through water (which, being a conductor, dissipates the electromagnetic field as electric current), whereas sound waves can do this and are thus more useful in underwater observations. Yet another factor concerns whether it is practicable to construct receivers and perhaps transmitters also for a particular information carrier at a reasonable cost, given environmental and other constraints. Measuring devices to be carried by wild animals or to be

placed into human bodies are limited in size and weight, so that a signal carrier requiring much energy or large aerials cannot be used. Similarly, radar aerials for aircraft are limited in size, which constrains the choice of wavelengths which may be employed.

2. Signal strength considerations: Because of distance, only a very small part of the energy given off by the data sources will reach the signal receivers. The signal will, in fact, be attenuated in an inverse ratio to the square of the distance from the source. Further, the signals are subject to attenuation by the intervening medium and to various kinds of interference. One is thus faced with the problem of how to maximise the signal-to-noise ratio so that useful information may be extracted. It is not sufficient to amplify received signals, since this increases the power of noise to the same extent as it does the signal.

3. Signal design: In both active remote sensing and telemetry there are limits to the power of the transmitter, and the signals transmitted have to be designed such that they can be recognised correctly despite the presence of noise.

All these considerations are interconnected, and developments affecting one consideration would often lead to reassessment regarding other factors also. The subject is still in a process of constant evolution. In the following sections we shall discuss briefly the basic principles underlying common remote data acquisition systems [1, 2, 3, 4].

8.2 Passive Remote Sensing

8.2.1 LIGHT

Visible light provides by far the most important means of passive remote sensing because of the factors listed below.

1. Availability: The Sun provides intense illumination of all things on Earth at no cost to man. To a lesser extent, so do the stars. (Moonlight is reflected sunlight.) Many objects also emit light, though this is usually insufficient for remote sensing purposes. However, in satellite pictures of the Earth at night, the light emitted by population centres provides an excellent map of human inhabitance.

2. Information carrying capacity: It is obvious enough that pictures of objects do provide much information about them. More importantly, in a remote sensing context, the frequency content of light derived from a data source is of major importance, since each material has its own characteristic way of reflecting or absorbing light, known as its characteristic **spectrum**. Consequently, analysis of frequency content is an extremely important method for extracting information.

3. Ease of detection: A variety of methods exist for light detection, not least our own visual sense. It is both cheap and fairly simple to record light emission from a wide area on photographic films, with colour films providing a rough indication of the frequency content of light from each source. By passing the light through different filters and recording each result separately, we can obtain a somewhat more accurate idea of frequency content of the incoming signal. More exact results may be obtained by the use of photodetectors and spectrometers or interferometers.

4. Signal strength enhancement: It is relatively easy to focus light received over a wide front by means of lenses and mirrors. This causes the signal energy to be accumulated while noise, being random, cancels itself out. Thus, telescopes with large entrance lenses produce sufficient accumulation of input signal such that even very faint distant stars may be observed. Another method of enhancing light energy is to take long exposures of faint light sources when recording light photographically. This also improves signal-to-noise ratio because, although the instantaneous value of the interference fluctuates, the average value over a long period of time is small and its main effect is to add a uniform 'background exposure' to the recording.

5. Resolution enhancement: By means of optical focusing one can also detect light selectively from a point source, so that it may be studied independently from its surroundings. This also helps to increase the signal-to-noise ratio since it excludes light from directions other than the point source, which would otherwise be included as unwanted information (noise).

    The main disadvantage of visible light as a source of information is its dependence on climatic factors, since light cannot easily penetrate clouds, rain or snowstorms, and is scattered by suspended dust particles.

We now look at two examples of visible light remote sensing.

Film recording from the air
Photographs taken from satellites or aircraft provide global views of large areas on the Earth's surface. By careful analysis from multiple photographs obtained in this way it is possible to discern its various properties. For example, the distinguishing features of grass, soil, concrete and bitumen can be determined from two photographs of a region taken in different frequency ranges. If we filter out frequencies outside the wavelength range 6000–7000 Å (here Å is an abbreviation for **Ångstrom unit**, where $1 \text{ Å} = 10^{-10}$ m), and record the picture on film sensitive to this particular range, then grass and bitumen will appear to be dark whilst concrete and soil will be shown as light areas. However, if the wavelength range selected is 7000–9000 Å, then grass and concrete will appear light and bitumen and soil dark. Consequently, any area which is light in both pictures may be assumed to be concrete, whereas common dark areas are likely to be bitumen. By making more elaborate differentiations of frequency bands, fairly detailed information about the terrain can be obtained, including identification of different soil types, vegetation, crop yield (since fields rich in a particular

harvest seed shows different colours from poorer crops) and humidity (through colour of soil or depth of colour of vegetation). Further details of this type of analysis, known as **texture analysis**, are given in other works [2, 5].

Out of many technical details which must be resolved to achieve success in this type of surveillance we shall mention two. The first is the question of **multi-band photo calibration**. Many photographs of areas having known ground features, whether rock, soil, vegetation or built-up zones, are taken at different frequency ranges. The appearance of each different type of feature under the various conditions and different observation frequencies is noted, with two aims. First, one chooses those frequency bands which are most useful for information extraction purposes. Second, the **frequency profile** of each type of feature is recorded. Later, when photographing areas having unknown properties in these frequency ranges we may compare the recordings obtained with the pre-recorded profiles determined earlier, so that information indicating the content of these areas and the conditions observed for each feature extracted can be obtained.

Next we consider extended exposures. Since the camera taking the pictures is located in the aircraft or satellite it will be moving relative to the ground and the image will not be static during an extended exposure. To ensure that light from the same source is always focused on the same spot on the film, it is necessary to traverse the film at an appropriate speed across the exposed film area in the direction opposite to that made by the vehicle carrying the camera. For example, when taking photographs at a scale of 1:100 000, forward movement of the vehicle by 100 m corresponds to a 1 mm movement of the image in the same period, so that traversing the film backwards at a steady rate by 1 mm whilst exposing the film would ensure the correct exposure conditions.

Optical astronomy

To observe distant and faint objects it is necessary to construct telescopes with very large entrance lenses (some have diameters of several hundred feet) and to focus all the light received onto a small detector area, thus achieving maximum signal enhancement. High resolution is attained since light from those stars removed from the optical axis does not fall on the detector. The detector may be the eye of an observer or a photographic plate. More frequently it is a spectrometer of some sort, since the frequency spectrum and polarisation of a star provide much information about its composition, surface temperature, mechanism of light production and object motion, as well as about the properties of the gas and dust clouds that intervene between the star and the telescope.

8.2.2 INFRA-RED RADIATION

As in the case of visible light, the availability of infra-red radiation is high, since all objects emit infra-red rays to varying degrees depending on their temperature. The rays provide useful information about both the composition of an object and its thermal state. While the principle of infra-red remote sensing is largely similar to that of light sensing, infra-red rays permit a smaller range of detection methods. Films sensitive to infra-red rays are available, but there is not the same colour—film range which we find with natural light. Hence the main use of infra-red films is in conjunction with normal photography, to provide an extra frequency band to the frequency profiles.

When observations are carried out only in the infra-red range, spectrometric techniques are usefully employed. The emissions from an individual data source are received by an interferometer or grating and the spectrum of the emissions is obtained in this way. Because the spectrum is determined by the chemical composition and the temperature of an object, much useful information may be extracted from it. However, it is relatively expensive to carry out simultaneous observations on an array of objects, since this would lead to duplication of the apparatus used many times. Instead, the same apparatus is used to observe different objects serially, i.e. one at a time, giving multi-band, single-object observations. Single-channel, multiple-object observation may be achieved by focusing light received from an area onto a rectangular array of photoelectric detectors, each recording the total energy from one point, but not capable of indicating the frequency content. Multi-channel, multiple-object observation by the use of **spectral imagers** has not yet been utilised on any large scale.

In a pilot study carried out over southern California, multiple-object observation, using three infra-red channels, was carried out by photographic means. It was found that emission from vegetation contains a high component of long-wavelength radiation (> 8000 Å), and that built-up areas show a preponderance of short-wavelength emissions (< 6000 Å), whereas deserts show up principally in the medium wavelengths (6000–7000 Å). Water shows up as dark zones, since water is a poor emitter in the infra-red range. This type of analysis enables terrain classification to be attempted in a way not easily carried out with visible light photography.

Another important use of infra-red recording is in the production of heat diagrams of factory plants, hospital patients or other heat-emitting objects. The 'brightness' of different parts of the object indicates the local temperature. In this way, infections, whether local or general, and tumours may be detected in patients, and heat losses, leakage of hot gas, or failure of insulation, may be observed in industrial plants.

A great advantage of infra-red radiation over visible light is the fact that objects *emit* it, rather than merely reflect it, so that observations may be carried out at night.

8.2.3   ULTRA-VIOLET AND OTHER RAYS

For various reasons, radiation utilising wavelengths shorter than that of visible light have only specialised applications in remote sensing. They are difficult to focus by optical methods. This is because they are sensitive to the imperfections in a surface which appears to be smooth as far as visible light is concerned, so that mirrors that reflect visible light well will not be so effective with rays of higher frequencies. (Generally, waves of any frequency are only affected by objects whose sizes are comparable to or larger than their wavelengths. This is why visible light is scattered by the water vapour particles in clouds, whereas radio waves are not.)

We are, however, witnessing a gradual transfer of techniques developed for laboratory experiments in physics or chemistry, such as ultra-violet spectroscopy, photomultipliers, fluorescence, Geiger counters, scintillation chambers, etc., to remote sensing applications where these special rays have particular advantages. For example, ultra-violet rays emitted by the Sun are heavily absorbed by the ozone layer before reaching the Earth. Measuring the level of this radiation thus gives an indication of the state of this layer, providing information about atmospheric pollutants, and surface activities on the Sun. X-rays have been used in hospitals since late in the last century for short-range observations (no focusing required), using fluorescent screens and special films. Their wavelengths are so short that they can, in effect, pass through the intermolecular gaps so as to penetrate through human bodies. They are absorbed by the various tissue material to different degrees, so that a picture may be formed of the interior arrangements of the body. Finally, as mentioned earlier, gamma-ray detectors may be used for discovering areas of terrain containing radioactive mineral deposits. Radioactive isotopes emitting gamma rays are also used as *trace* agents, e.g. to permit the observation of food or chemical absorption by a human body.

8.2.4 RADIO WAVES

The most important application of radio waves in passive remote sensing is in radio astronomy where they have sometimes decisive advantages in all-weather observation compared with their optical counterparts. Radio telescopes are constructed by placing individual aerials over a wide area and connecting them together in such a way that radiation from one particular direction is additive whereas that from other directions cancels. This is analogous to the focusing power of optical telescopes. As the Earth rotates, the aerial array is directed at different parts of the sky, and radio maps may be plotted to pinpoint sources of high radiation. Focusing can be achieved adequately only if the overall size of the telescope is considerably larger than the wavelength detected. In consequence, radio wave observations of objects on Earth are possible only in the microwave range, for wavelengths of the order of centimetres or less, so that receiving aerials of manageable sizes may be employed.

For use in passive remote sensing devices the use of wavelengths in the microwave range present two additional difficulties: the signals are detected at a low energy level and it is not possible to expect high reproduction of details. While objects emitting infra-red rays always emit some energy in lower frequencies as well (the amount at each frequency being in accordance with the characteristic spectrum of each object), the energy emitted is only a very small portion of the total power. Also, for reasons explained earlier, structures smaller than their radiated wavelength cannot be 'seen'. Finally, the detected signal is subject to a great deal of manmade interference, ranging from broadcasting stations to electric machinery. For these and other reasons we may regard passive microwave sensing as being still in an early stage of development.

8.2.5   VIBRATION

A long-established remote sensing process using mechanical vibration (shock waves) is that of seismic observation. One example is earthquake monitoring, in which a number of seismic detectors are placed at various locations and their readings recorded, usually at a central monitoring station, although local recordings may also be made to provide a back-up source of information in case of data transmission loss. Shock waves generated by the movement of geological faults or volcanic eruptions are received and recorded. Several types of information may be extracted from these recordings. The shapes of the received pulse trains provide indications of the type of seismic activities responsible. By comparing the arrival times at various observation stations, which are determined by the distance of the shock source from each station, one can pinpoint the location of the source. Once this has been determined it is also possible to estimate the strength of the activity by studying the size of the shock pulses, taking into account how far they have to travel before reaching each station.

Seismic detectors are also used to monitor the stress and strain in manmade structures such as dykes and dams or roads and airport runways. As a reservoir is filled, increasing pressure is exerted on the dam, and minor settling may occur, especially in a new construction. By monitoring such activities continuously, advanced warning is obtained in case of serious failure; a similar situation holds for roads and runways. As vehicles travel over them, mechanical vibrations are generated; these may be recorded for such purposes as analysing the density of traffic on a country highway or detecting any weakness in the foundation to the structure.

8.3   Active Remote Sensing

8.3.1   RADAR

Microwave radar systems constitute the most important means of active remote sensing. Beamed transmission signals are easy to generate by the use of electromagnetic oscillators tuned to particular frequencies. The technology of emitting and focusing a microwave beam using multipole aerials is well developed so that a concentrated beam may be directed at any object under observation. Microwave signals continue to have good air penetration capability under any weather conditions. It is also possible to use special carrier modulation techniques which impose certain modulation patterns [6] on the transmissions which are easy to recognise; thus, reflections from the objects under observation can be identified despite noise and distortion. An additional advantage arising from the use of microwave transmissions is that it is possible to estimate the speed of a moving target by observing the frequency shift it causes on the reflected waves. This is known as the **Doppler effect**.

The use of radar in aircraft tracking is now well known. A rotating aerial emits a pulse train having a prescribed pattern, imposed on a wave of some fixed frequency. The beam rotates or sweeps from side to side to cover the area of interest. If a pulsed beam strikes any object, a reflected signal is generated and detected by the receiving aerial. The distance of the object or target from the aerial may be estimated from the time taken for the pulse train to travel to the target and back, and its direction is known from the direction the aerial is pointing towards when the return signal is received. The aerial may have rotated by a small amount during the travel time of the pulse but this may be neglected because of the very high speed of wave travel. Knowing its distance and direction, the position of the target can be determined by displaying the reflected signals automatically on a CRT screen, although slower hand calculations may also be made. The direction of travel and speed for the target may be estimated from successive positions and elapsed time between them as observed during each rotation of the aerial. Alternatively, this can be obtained from measurement of the Doppler shift, which may be more accurate, particularly at directions directly towards or away from the transmitting aerial.

An estimate of the size of the target may also be made from the amount of energy it reflects, making due allowance for the fact that the radiation is subject to attenuation over distance in inverse ratio to the square of the distance travelled.

A recent application of radar techniques is in rainful estimation. A wavelength in the millimeter range is used so that the microwave signal can be reflected from individual raindrops. The greater the size of the raindrops and the more dense their conglomeration, the greater is the amount of reflection, again allowing for distance. As in the case of aircraft tracking, a rotating aerial is used and reflections from all directions are continuously recorded. However, it is a little more difficult to recognise the return signals because they arrive over a continuous period of time, from the nearer to the more distant parts of the area of rainfall. However, it is feasible to accumulate the amplitude of the returned signal, weighted by a factor that decreases with distance, to arrive at the total amount of rain in a particular direction. The area of the rainfall can be plotted from the duration of the returned pulse train.

A different way of employing radar is the technique of **side-looking radar** used for mapping overflown terrain from the air. This is illustrated in Figure 8.1. An aerial carried by the aircraft is designed to emit a 'fan' of microwave pulses towards one side of the aircraft which are reflected from a series of targets at ground level situated on a straight line perpendicular to the path of the aircraft. Because each target is at a different distance from the aircraft return signals are received at different times. Processing of this series of return signals will permit identification of individual parts of the line of targets. As the aircraft flies forward its 'fan' of microwave signals sweeps over an area in which the targets are situated (Fig. 8.2). This allows the return signals to be plotted for the entire area as a two-dimensional figure.

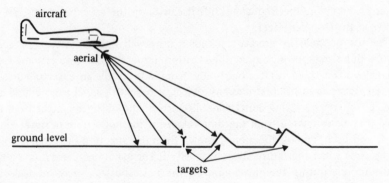

Fig. 8.1   Side-looking radar

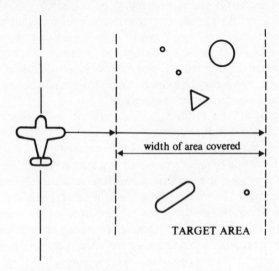

Fig. 8.2   Sweep of side-looking radar

Although side-looking radar is also useful for other purposes, its application to mapping is most important because of a special factor. Terrain that slopes towards the aircraft with its ground surface facing it, reflects a higher proportion of the signal to the aerial than surfaces which slope away from it (see Fig. 8.1). Thus, one slope would show up as a 'bright' area on the plot, and the other a 'dark' area. Consequently, side-looking radar provides a simple method for obtaining a map of a large area of terrain, including important contour levels, very quickly.

Both the rotating and the side-looking radar aerial are longer in the horizontal dimension than vertical. This is to achieve high focusing horizontally, so that the beam has an unambiguous direction, but fans out over a range of vertical directions so that multiple targets in the same plane may be observed. As mentioned earlier, to achieve focusing, the size of the aerial must be much greater than the wavelength of the object to be detected.

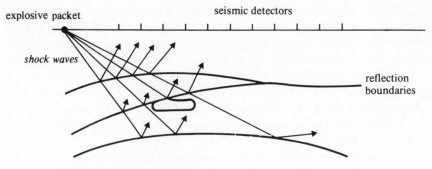

explosive packet   seismic detectors

shock waves

reflection
boundaries

Fig. 8.3   Active seismic exploration

8.3.2 SEISMIC WAVES

Active remote sensing using seismic waves is common in mineral exploration and
geological surveys. Shock waves are generated by exploding packets of dynamite
in holes drilled into the ground to some particular depth, and reflected waves are
measured with an array of seismic detectors (Fig. 8.3). The reflections occur
at boundaries between different geological formations, such as between soil and
underlying rock, or at horizons containing underground water or oil. Reflections
reach detectors at various times, depending on the position of the boundaries in
relation to each detector. By recording the reception at each detector and analysing the arrival times of the shock-wave reflection and reverberations, one can
obtain a good idea of the depth of each layer of geological material as well as its
composition. The latter is possible because the speed of seismic waves varies with
the layer density.

Seismic arrays are also used for monitoring the magnitude and direction of
seismic disturbances and earthquakes, as mentioned earlier. By including a phased
series of delays with a different delay value for each seismic recording channel,
it is possible to sum the array signals in such a way as to augment the signal
arriving from a particular direction relative to the line of the array. This is
because the individual delays for each channel are directly related to the position
of each seismometer in the array. By adjusting the set of delays in a certain way
over a period of time, noting the value of the set which gives rise to the maximum signal, the direction of the seismic disturbance may be determined.

8.3.3 SONAR

Directional echo-sounding with water as the intervening medium instead of air or
space is carried out with low-frequency compressional waves having frequencies
ranging from the high audio frequencies up to a few hundred kilohertz. Techniques similar to those of radar are employed, but the method is termed **sonar** in
deference to its sound wave origin. The significant advantage of sonar waves over
microwaves is of course their low attenuation when travelling through water
where microwaves would be rapidly dissipated. They are usually generated using
a crystal oscillator, which employs a thin slice of quartz or other material situated

in a rapidly alternating electric field. Variations in the field strength cause the material to expand and contract at a high audio frequency. It is possible to focus these waves using reflectors. When, however, a high-power beam is required, an array of individual oscillators will be used. By feeding the same electric oscillation to the whole array, the sound waves emitted by the individual oscillators are in phase, and the total emission is highly directional, since the emissions are in phase addition along a direction perpendicular to the array, but tend to cancel in other directions. This is similar to the focusing capability of a multipole radio-wave aerial array.

The use of sonar in ship and submarine tracking is similar to that of radar. Because mechanical vibrations are present almost universally, reflections can be recognised only if they are reasonably strong. Consequently, the operational range of sonar is much less than that of aircraft-tracking radar. However, because of the lower cost of individual sonar receivers, compared with a microwave system, it is feasible to record waves received over a whole front by the use of an array of separate aerials and detectors. (In contrast, a microwave receiving aerial usually focuses the incoming wave so that it is received by only one detector.) This enables a more complex information extraction process to be implemented, and this is similar to the system for processing seismic array signals we mentioned earlier. A recent development is underwater picture reconstruction using sonar, a process illustrated in Figure 8.4. Here sound waves reflected from a target are recorded by an array of microphones, of which only three are shown. The electrical signals derived from the microphones are each subject to a given phase

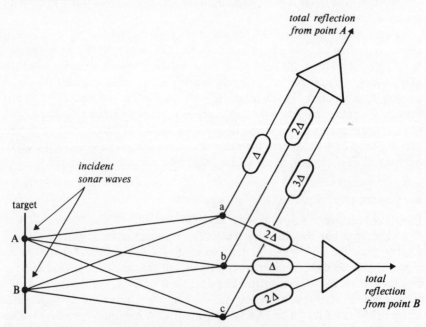

Fig. 8.4   Picture reconstruction using sonar

delay, Δ, and then added. The phase delay is related to the length of transmission path, so that, considering signals reflected from point B, a longer delay is required for paths Ba and Bc than for the more direct path Bb, in order that the reflected signals at all three detectors are summed in phase. Signals reflected from other areas in the target will add differently, since the delays we have postulated are appropriate only to point B. In general, signals arriving from all other points would tend to sum to zero and contribute little to the total summation. If we consider the appropriate addition for another point, A, then an entirely different set of delays is required for the three channels due to the different transmission path lengths. In a practical situation, a larger number of microphones would be used arranged in a linear or matrix array. The phase delays associated with the array would be arranged to provide a maximum response from the summation of all the microphone outputs for each target pixel. This could be carried out in parallel, in which case we obtain a picture of the target 'illuminated' by sound waves in place of the more usual light illumination. The principle of this system is similar to that of a lens or a telescope, which focuses light coming along each different direction on to a different point on the image plane. Alternatively, a serial operation is carried out with the set of delays programmed to give outputs for each target pixel in turn as the picture is effectively 'scanned' by the delay value adjustment. Whilst the parallel process may be implemented either by analog or by digital means, the serial process requires a controlling computer or microprocessor to effect the necessary sequential delay adjustment.

8.3.4 LASER

Until recently, visible light was little used in active remote sensing because of the need for a very high-power light source and the presence of extraneous light causing difficulties in extracting the reflections from the received signal. The laser light source overcomes these problems in quite a remarkable way. Ordinary white light is said to be **incoherent** − that is, its energy is distributed in all directions and over a wide frequency band. Further, its separate frequency components are not in phase with each other. In fact, it behaves in much the same way as noise in electrical communication systems. Hence the term 'white noise' which is widely used to describe this kind of chaotic transmission property. In order to produce a narrow beam of light having significant energy we need to focus the light with quite large parabolic reflectors. The light is then said to be **collimated**. The significant property of the laser beam is that it is *produced* as a collimated beam and is composed of regular and continuous waves of a single frequency and constant amplitude. Hence, a laser system can emit a highly focused and narrow light beam of great intensity, pure frequency and uniform phase. This means we can use a narrow band filter to remove all other frequencies that might also be received, so that the amount of interference is greatly reduced, allowing the detection of very small reflections.

All the signal recording processes used in passive remote sensing may also be employed here. However, multi-channel studies require the generation of light at several different frequencies, calling for either a number of different lasers, or more economically a tunable laser whose frequency of light emission is variable. The receiver will require a filter with the appropriate frequency characteristics, either having passband that is just broad enough to cover the whole frequency range of the tunable laser, or having a narrower passband and an adjustable centre frequency that changes in step with the emission frequency of the laser.

Since a light source is used to illuminate the targets to be observed, it is not necessary to limit observations to daylight periods. Further, a frequency can be chosen particularly suited to a given requirement. For example, in one experiment a ruby laser (wavelength = 3472 Å) was used for monitoring the content of water vapour in the atmosphere because of the particular scattering effect of water molecules observed at that frequency. Another application feasible with laser techniques is the measurement of target speed through the Doppler effect, but this can be employed only if the light being reflected is generated at one specific frequency.

8.4 Introduction to Telecommunication

8.4.1 DIRECT TRANSMISSION OF SIGNALS

In this section we discuss the basic principles of telecommunication as applied to the transmission of signals. In realistic circumstances, it is seldom feasible to implement the direct transmission of measured signals from the transducer to the remote data system, using a pair of wires to send the transducer output to the system. The first difficulty experienced with such a simplified system is the cost of setting up and maintaining the communication lines which would be extremely high for distances greater than a few hundred metres, even if it were technically feasible. Only a public utility, with many customers to share high-cost facilities, can afford such systems. In addition, signals propagating over such lines are subjected to attenuation as well as interference so that the values received would be unreliable without recourse to amplification along the route.

Both sources of inaccuracy, as one would expect, increase with distance. Direct transmission is also subject to the physical characteristics of the transmission line in a very fundamental way. A parallel pair of electric wires spaced in some insulating medium behaves as an electrical capacitor. This, combined with the inductance of the long conductors and their resistive value, causes the line to have frequency-dependent characteristics. When a pulse is sent through the line it does not retain its original shape during propagation as the capacitance causes the energy contained in the pulse to be stored and gradually released. Thus a short pulse is stretched to become an extended one, and a long pulse is distorted at the leading and trailing edges. Severe forms of this distortion can actually cancel out subsequent peaks. The effect, besides causing distortions, also limits the rate of transmission, since we must send successive signal pulses

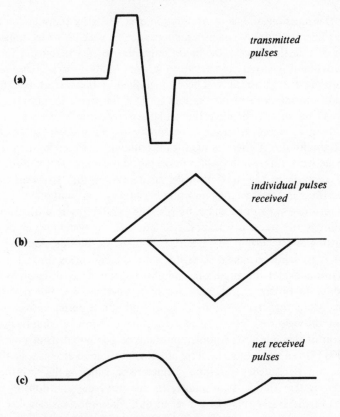

(a)   *transmitted pulses*

(b)   *individual pulses received*

(c)   *net received pulses*

Fig. 8.5   Pulse stretching effect of low-pass channel

at sufficiently well-separated intervals so that they may continue to be received distinctly at the far end.

Mathematically, the pulse stretching effect is equivalent to low-pass filtering. If an attempt is made to transmit a high-frequency signal through such a medium, very little signal will reach the receiver, since the stretching effect causes cancellation between successive peaks and troughs. The higher the frequency, the more severe is this effect. (See Fig. 8.5 for a graphical illustration.)

The above discussion also shows that there is a relation between the data transmission capacity and the frequency limitation of a communication line caused primarily by line capacitance. Generally speaking, a communication line with a bandwidth of f Hz can transmit not more then 2f pulses per second – a result we would expect from the sampling theorem (Section 5.2). Higher quality cables generally have a higher frequency cut-off value and produce less pulse shape distortion. In particular, coaxial cable, which has a central inner conductor surrounded by a concentric outer conductor, gives lower distortion and is also less susceptible to interference, since the outer conductor can be earthed to shield the inner conductor (along which the pulse travels) from external induced interference.

The transmission of analog information over a lengthy transmission cable will require some preliminary channel calibration; this is achieved by transmitting a number of trial signals over the line. Comparison of the shape and form of the transmitted and received pulses will show what kind of distortions and interferences are present. The latter will be distinguished by their effect in adding to the signal and causing random fluctuations in it. If the transmission tests show that the system produces little distortion and interference, and merely attenuates the pulses, then it is only necessary to amplify the received signal to restore its original amplitude. If there is significant random interference, then it is necessary to design a filter which will remove the added noise. This will require a frequency analysis of the noise in order to determine the characteristics of the required filter and may be done in a number of ways, as mentioned in Chapter 3.

The restoration of pulse shape by filtering is rather more difficult. One must first estimate the frequency response function of the system which includes the transmitting and receiving equipment, as well as the transmission line. A direct method is to send sinusoidal signals of various frequencies through the system and record and plot the amplitude and phase of the received signals. An indirect method is to obtain the impulse response function, i.e. the received signal resulting from the transmission of short and sharp pulse, known as a **delta-function** through the system. The frequency response is simply the Fourier transform of this impulse response, and may be computed from a record of the latter [7]. Given the frequency response, it is possible to attempt the design of a filter having a frequency response characteristic which is the reciprocal of the measured system frequency response. In this way, the variations in frequency response of the system may be cancelled out. The process is called **equalisation** by communication engineers. In actual fact, there are two serious difficulties preventing fully successful equalisation from being carried out. First, a given frequency response may be physically unrealisable with a practical filter design. Further, since the system attenuates high frequencies, the equalising filter must amplify them, and this causes a deterioration of the signal-to-noise ratio due to the predominance of high frequencies contained in the noise accompanying the signal. Thus, the objectives of equalisation and noise elimination are conflicting, and a compromise has to be reached between them. It is now usual to find in communication equipment a number of frequency-variable filters associated with the transmission input—output lines which can be adjusted to reach this compromise. These are termed 'equalisation controls' and are set up in conjunction with specially designed line measuring test equipment which produces and measures the special test waveforms required for this purpose.

In the preceding discussion we have been concerned about preserving the 'shape' of the transmitted signal, i.e. its analog characteristics. The direct transmission of digital information is easier because accuracy of amplitude and shape is less important. As long as the received signals permit us to distinguish between 0 and 1 correctly then no error will be generated. However, timing becomes more critical, since the receiving end must be able to recognise where one digit ends and the next starts. Usually the transmitter and receiver have timing

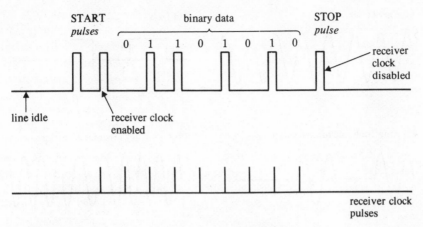

Fig. 8.6   Asynchronous method for byte transmission

generators or 'clocks' set at the same rate, so that the latter will accept bits at the same rate as the transmitter sends them. It is also necessary to formulate some mechanism for grouping bits into individual numbers (words or bytes), since the significance of an individual bit depends on its position within a word.

A simple method for sending binary data is the **asynchronous method**. Data are sent in eight-bit units (bytes), preceded by two START pulses and followed by one STOP pulse (see Fig. 8.6). They are spaced out at prescribed intervals, say 1/300 s. The first START pulse initiates the timing or clock pulse generator of the receiver, which will produce pulses at the rate of 300 per second. The second clock pulse coincides with the arrival of the second START pulse. If both are received together then this indicates that transmission is proceeding correctly, and the next eight pulses are accepted as the transmitted data by opening the input path with each clock pulse. The last pulse, the STOP bit, halts the receiver clock generator, which will remain inactive until it is restarted by the START pulse of the next transmitted byte. The method is simple, though somewhat slow, since it requires the transmission of pulses carrying no information and a sufficiently slow transmission rate to allow the receiver time to start and stop its clock pulse generator. **Synchronous methods** are more involved and we refer the reader to texts on communication for details [6, 8].

8.4.2   CARRIERS AND MODULATION

In most telemetry applications, signals are transmitted by means of carriers. A **carrier** is a high-frequency signal conveyed over transmission lines or transmitted by radio and capable of being modulated by the signal or message it is desired to transmit. This latter can be at a low transmission rate or low frequency so that the carrier system will enable the small bandwidth required to convey the information to be translated into a higher frequency band which is easier to convey through communication equipment. For short-distance mobile or space communication it is common to employ radio transmitters and receivers. Long-range

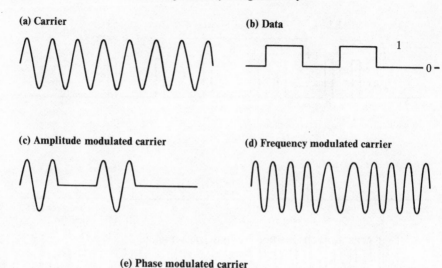

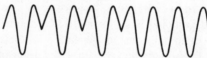

Fig. 8.7   Carrier modulation methods

ground transmission is normally carried out via the public telephone network. In each case, the system uses special high-frequency carrier signals that can be generated, transmitted and received through the system conveniently and economically. The information contained in the data is imposed upon a carrier, to produce a signal that satisfies both the needs of the system as well as those of the user. This process is called modulation. At the receiving end the combined signal is demodulated to recover the original information. There are three basic mechanisms for modulating a sinusoidal signal carrier: by imposing on its amplitude, frequency or phase. These are illustrated in Figure 8.7, where the carrier is shown in (a) and the modulating signal as the two-level waveform in (b). Amplitude modulation is the simplest form of modulation and represents the product of the two waveforms as shown in (c). In frequency modulation, the two levels of the modulating waveform are translated into two different but constant carrier frequencies, as in (d). Phase modulation, shown in (e), leaves the frequency of the carrier unaltered but does affect the phase of the carrier compared with the unmodulated waveform. This is shown as an 180° shift occurring at each level change of the modulating waveform.

Demodulation in the case of an amplitude-modulated carrier involves taking the absolute value of the composite signal and then subjecting it to low-pass filtering to remove the high-frequency (carrier) components. This is illustrated in Figure 8.8. Frequency and phase demodulation are rather more complicated and cannot be usefully discussed here [8].

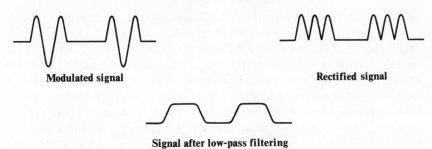

Modulated signal   Rectified signal

Signal after low-pass filtering

Fig. 8.8   Amplitude demodulation

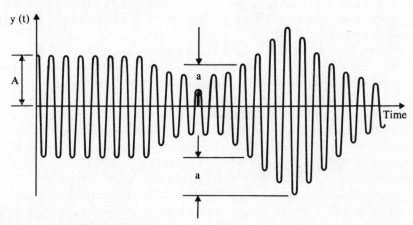

Fig. 8.9   Amplitude modulation

What effect do these various forms of modulation have on the transmitted frequency bandwidth? Let us look at the situation with amplitude modulation. If the amplitude of a sinusoidal carrier of frequency f_c is varied sinusoidally between the limits $A \pm a$, as shown in Figure 8.9, with a modulating frequency, f_m, considerably lower than that of the carrier frequency, then the equation of the resultant waveform will be

$$y = (A + a \cdot \sin \omega_m t) \sin \omega_c t$$

$$= A(1 + a/A \cdot \sin \omega_m t) \sin \omega_c t \qquad (8.1)$$

where

$$\omega_m = 2\pi f_m \quad \text{and} \quad \omega_c = 2\pi f_c$$

In equation (8.1), a/A is known as the **modulation factor** and $100a/A$ as the **percentage modulation**. This modulated waveform does not consist simply of frequencies f_c and f_m, as we shall see if we expand equation (8.1)

$$A(1 + a/A \cdot \sin \omega_m t) \sin \omega_c t = A \cdot \sin \omega_c t + a/2 \cdot \cos (\omega_c - \omega_m)t$$

$$- a/2 \cdot \cos (\omega_c + \omega_m)t \qquad (8.2)$$

Here the term $A \cdot \sin \omega_c t$ corresponds to a sinusoidal waveform of the carrier frequency f_c of maximum amplitude A. The second term, $a/2 \cdot \cos(\omega_c - \omega_m)t$, corresponds to a sine (or cosine) waveform of frequency $(f_c - f_m)$, of maximum amplitude $a/2$. The third term, $a/2 \cdot \cos(\omega_c + \omega_m)t$, corresponds to a sine (or cosine) waveform of frequency $(f_c - f_m)$ of maximum amplitude $a/2$. These latter two terms are known as the **sidebands** of the modulated carrier. We see from this that amplitude modulation broadens the bandwidth of the unmodulated carrier frequency so that the bandwidth now becomes $2 \cdot f_m$ Hz wide. A similar but more complex analysis shows that frequency modulation produces a much more drastic broadening effect, whilst that of phase modulation also increases the bandwidth of the original unmodulated carrier waveform.

In digital transmission, we are concerned with just two levels of the modulating signal. Analysis will show that this too modifies the bandwidth of the carrier frequency.

If we were to perform spectral analysis on the composite signal formed by amplitude modulation of a carrier by a two-level signal (Fig. 8.7c), we would find that its spectrum would be identical with that of the original modulating data of Figure 8.7b, except that the centre frequency would have been shifted to the carrier frequency. In addition, frequency and phase modulation will cause a further increase in bandwidth. In all cases, the following statement is true: one cannot *reduce* the bandwidth requirement by modulation. If the original data consist of so many pulses per second, a line with a specified frequency limit will be required to transmit them directly. Alternatively, if we use the data to modulate a carrier, then we will need a different sort of system that can transmit the carrier efficiently. However, since we need to demodulate the carrier at the receiving end and recover the original data correctly, then the system must have a bandwidth at least equal to that required for direct transmission. Usually, the bandwidth requirement is increased. This is a basic limitation of communication systems. A communication system with a given bandwidth and signal-to-noise ratio has a maximum rate of reliable information transmission. Any attempt to exceed that limit, regardless of the data transformation processes employed, will only lead to increased error rates.

For example, suppose we have a system capable of transmitting P pulses per second, with pulse height ranging from 0 to V. We can impose onto each pulse several bits of information. That is, we represent an n-bit string $x_1 x_2 \ldots x_n$ by a pulse of height $xV/2^n$, x being a binary number consisting of the bits x_1 to x_n. This appears to increase the data rate by a factor of n, as each pulse carries n bits, rather than just one bit of information. However, the likelihood that a pulse of height $xV/2^n$ might be received erroneously as $(x \pm 1)V/2^n$ is quite high. Consequently, the error rate must also increase. Only if the communication channel is very noise-free, would the increase be negligible. We see that the rate of information tranmission achievable on a channel depends on the signal-to-noise ratio as well as the bandwidth, so that it is possible to reach a compromise between error rates and information transmission rates within certain limits.

An example of such a compromise is the method of pulse code modulation. Here, analog information is transmitted in digital form as a series of binary pulses, which are reconstituted into analog form by the receiver. In such a system, one sends n binary pulses in place of one analog pulse, increasing the bandwidth requirement by a factor of n. However, the probability of error is considerably reduced, because one only has to be able to distinguish between a 0 pulse and a 1 pulse. Further, it is not essential to ensure that the system will neither amplify nor attenuate the pulses, since the exact amplitude of each pulse does not matter. Such methods are now being increasingly implemented in telephone systems, high-quality tape recorders and even musical discs.

Another general observation we wish to make is that none of the three modulation methods really lends itself well to the transmission of analog data. In both amplitude and phase modulation, it is difficult to recover the original values accurately. In the former case, difficulty arises because any attenuation suffered by the composite signal is passed on to the amplitude of the signal when it is recovered, and one cannot reverse this in normal circumstances because the attenuation factor is variable and unknown. In the latter case, phases of sinusoids are affected by time delays, which are again variable and uncertain; also, phase demodulation processes themselves are not accurate for analog information. On the other hand, frequency modulation is accurate and little affected by these factors. However, it is complex and its drastic bandwidth broadening effect makes it unsuitable for the simple equipment which we are obliged to use in telemetry. Generally speaking, it is desirable to digitise analog information for use with carrier systems, and this is the direction towards which most carrier systems are now progressing.

8.4.3 MODEMS

A special form of modulation—demodulation system has been developed for the transmission of digital data over telephone lines. This is widely used to transmit computer information to remotely connected terminals and to link computers together. The unit that carries out the process of conversion of digital data into a modulated form suitable for transmission over telephone lines and the reverse process from modulated data to a digital data stream is known as a **modem** (modulator—demodulator). It is essentially an interface between telephone line and the computer equipment as shown in Figure 8.10a. A modem is required at both ends of a transmission link. A communication channel may be able to carry more data than one terminal can produce. It is possible to use this spare capacity by connecting several terminals to the same line as shown in Figure 8.10b. This is known as a **multi-point system** and, as shown, indicates that each terminal has its own input connection to the modem and that the modem connected to the computer has a number of separate outlets, one for each of the terminals connected to the input modem. Other systems employing modems enable a number of terminals to be connected to a single computer input and, in this case, transmission to and from the terminals must include some means of identifying the source and destination of a particular data message. Such a system is called a **multi-drop line**. The reader can refer to the many books on data communication,

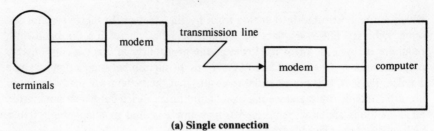

(a) Single connection

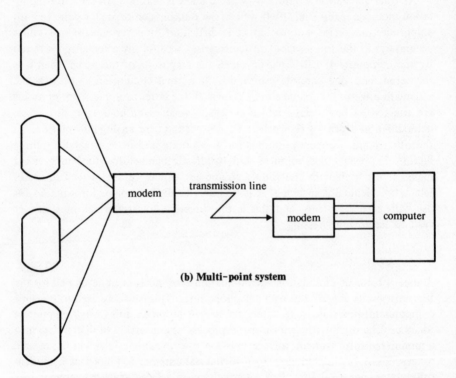

(b) Multi-point system

Fig. 8.10   Use of modems in communication with a computer

some of which are listed at the end of this chapter for further information [6, 8–10].

The practical limit to data transmission rate over telephone lines is reached at 9600 bits/s. Above this rate the accuracy of transmission is low and, although speeds up to 48 000 bits/s are in use, they require special transmission circuits. As far as the users are concerned, it is the choice of modem which actually dictates the speed of transmission and the transmission link is usually designated to operate at specific fixed speeds of 110, 300, 1200, 2400, 4800, 9600 and (using special lines) 24 000 and 48 000 bits/s. A modem operating at a given speed must include the appropriate equalisation and line-conditioning equipment for that speed.

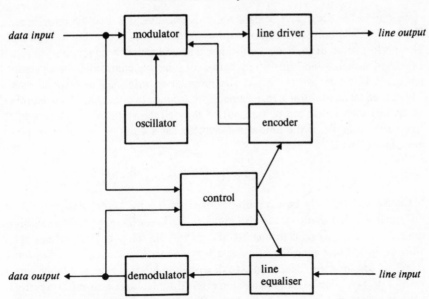

Fig. 8.11   Schematic diagram of a modem

A basic modem configuration is shown in Figure 8.11. High-speed modems would also include line-conditioning filters to compensate for the attenuation and delay distortion encountered in the telephone network. The encoding or modulation method can be any of the modulation methods discussed earlier. Amplitude modulation is not generally used except at very low speeds, because it is most likely to cause data corruption. At the lower end of the speed range, frequency modulation is most commonly found. This is often referred to as **frequency-shift keying** since, in fact, only two modulation frequencies are used, corresponding to a digital 1 and a digital 0. Phase-shift modulation (or keying) is universally used at higher transmission speeds as it offers the best degree of immunity against data corruption.

A third factor in defining the properties of a modem is its *mode* of operation. Asynchronous and sychronous modes of operation have been described earlier. Both are used in modem interface equipment, and in some units a switch is included to change from one mode to another. This is essential where, for example, the modem offers a wide range of working speeds, since phase-shift keying can be used only in synchronous mode and this modulation method, it will be remembered, is used for the higher speeds.

A switch may also be included in the modem to select either **full-duplex** or **half-duplex** mode of operation. In half-duplex mode a terminal can send or receive data, but not simultaneously — transmission can only take place in one direction at a time. In full-duplex mode the terminal can both receive and transmit data at the same time, i.e. in both directions. As would be expected, full-duplex operation requires a four-wire connection compared with the simple arrangement of two wires for half-duplex connection. Nevertheless, full-duplex connection is the most common arrangement found in present-day equipment.

In our earlier discussion of data transmission rates, we used the term 'bits/s'. However, this is only one way of expressing the rate of conveying information. The performance of modems (and other data transmission equipment) is often expressed in **bauds**, and this often causes considerable confusion when attempts are made to relate this to bits/s. The modulation rate may be given in bauds which is equal to one unit signal element per second. In effect, this is the number of distinct pulses that can be transmitted through the system during a second of time. For example, if the unit signal element has a duration of 20 ms then the modulation rate is

$$\frac{1}{0.02} = 50 \text{ bauds}$$

If the pulses have only two amplitudes, then each pulse carries only one bit of information and the bit rate is also equal to 50. However, at high data rates it is usual to transmit more than one bit of data with each signal. For modems operating at 4800 bits/s then the bits/s rate is *not* equal to the baud rate. We should therefore use the term 'bauds' only to describe **modulation rate**, and bits/s where we wish to talk about **data transmission rate**. Although data transmission systems are not necessarily based on individual pulses but employ more complex forms of signals to transmit information, the mathematical relation between the baud rate and the bit rate is generally present.

8.4.4 CHANNEL UTILISATION – MULTIPLEXING AND ERROR PROTECTION

We said in the previous subsection that the ability of a channel to transmit information increases with its bandwidth. When sending data whose data rates do not fully utilise the bandwidth of the system one has a choice: either the spare capacity may be used to provide error protection to achieve better accuracy, or several streams of data may be made to share the same channel to reduce costs. Obviously, the first option would be more desirable if the channel were noisy, while the second would be chosen if the opposite were the case. In many designs, both options may be implemented in part. The two options are called **error protection** and **multiplexing**, respectively.

Carriers provide one multiplexing method. The frequency range of a high bandwidth channel is divided into a number of sub-ranges, the width of each sub-range matching the requirement of a single data stream. A sinusoidal waveform having the centre frequency of each sub-range is used as a carrier for the data. The individual carriers, each modulated by its data stream, are added together (Fig. 8.12). Because the data streams have limited bandwidth compared with the carrier frequency, the frequency content of each modulated carrier is confined within its sub-range: therefore the carriers do not interfere with one another and may later be separated, using filters whose passbands coincide with the sub-ranges. Each filter selects an individual modulated carrier, which is then demodulated to recover its data stream. This process of **frequency division multiplexing** is commonly employed in telephone systems to link the

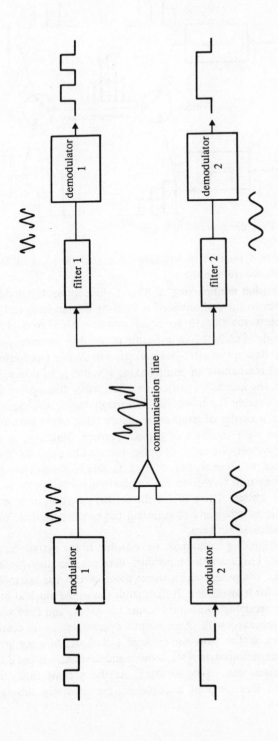

Fig. 8.12   Frequency division multiplexing

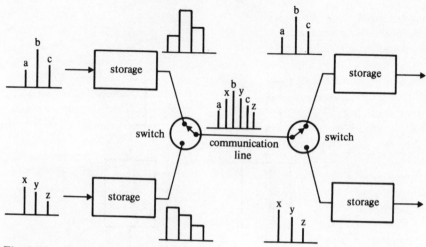

Fig. 8.13   Time division multiplexing

transmission of a number of low-rate lines, such as local substations, to high-rate, long-distance trunk routes.

In **time division multiplexing**, a cyclic switch connects individual pulse transmitters in turn to the communication line. At the receiving end, a synchronised switch connects the line to an equal number of receivers (Fig. 8.13). Thus, during one n-th of each period, a specific transmitter—receiver pair is connected. This allows n slow transmitter—receiver pairs to share a fast communication line.

The actual mechanism of multiplexing systems is of course more complex. For example, the individual carriers of a frequency division multiplexing system may not be suitable for transmission through the system because the system operates with a carrier of another frequency. One might have to modulate the system carrier with the sum of the sub-carriers. Similarly, in a time division multiplexing system, the series of pulses received through the cyclic switch may be transformed to other shapes, or used as data to modulate a carrier which is demodulated in turn to recover the original pulses before these are sent to the receiver cyclic switch. There is also the further question of how to synchronise the two cyclic switches and to maintain the synchronisation. We will not consider such system details here.

Other multiplexing techniques are possible if the system incorporates computer control. For example, individual data sources may present their transmission to the computer, which stores them until it has assembled a complete block of data for transmission. It then sends the whole block at high speed down the line. The receiving computer accepts the block, and then sends its content gradually, intermixed with other output operations, to its destination device. This is known as the **store-and-forward** technique, offering great operational flexibility since various data rates, sources and destinations and different management mechanisms can all be effected. At the present time, the methods of frequency and time division multiplexing are relatively simple, and may be implemented using inexpensive electronic components, though with low-cost microprocessors and memory chips becoming commonly available the situation may well change in the near future.

We now turn to the question of error protection. Generally speaking, all error-protection schemes require the slowing down of the rate of transmission by adding to the data additional information derived from the original data using some mathematical formula. At the receiving end, the same formula is applied to the received data to check that the different parts of the data continue to obey the same relationships. If there have been some transmission errors, then the different parts are likely to be inconsistent, thus indicating errors in the transmission. If the formula has been well designed, it may even be possible to identify the cause of the inconsistency, i.e. the error in the data, which may then be corrected.

In computer data transmission, error detection, followed by re-transmission, is commonly employed. However, this disrupts the steady transmission of data, and also requires data storage to hold the data for re-transmission. Program control is necessary because of the complexity of the process, and a form of communication protocol must be used in order to permit correct action to be taken. This is because the transmission consists of a range of information: data, error-protection information, requests for re-transmission, acknowledgement of data correctly received and not requiring further storage, etc. Each type of information needs to be properly identified, and a pre-defined procedure must be followed rigidly to ensure that the two ends of the communication process act in step with each other. Communication protocols, of which this is one example, take many forms and are too extensive to be considered fully here. Some detailed information on their design and operation is given in the references at the end of this chapter [9–11].

For telemetry purposes, a simpler system of error-correcting codes is used which does not involve the elaboration of re-transmission and protocol procedures. Error-correcting codes, though mathematically more complex than error-detecting codes, offer several *operationally* simple mechanisms suitable for hardware implementation. An example is the **Hamming code**.

To understand this, consider the following example. Given four bits of data, abcd, we produce from them three check bits, xyz, with

$$x = a \oplus b \oplus d, \quad y = a \oplus c \oplus d, \quad z = b \oplus c \oplus d \qquad (8.3)$$

where $\oplus$ indicates an EXCLUSIVE-OR operation and detects if the expression contains an odd or even number of ones (see Section 1.6.1). Now, a appears in the first two equations, b in the first and third, c in the last two, and d in all three. If any one bit is received incorrectly (assuming no other errors are present), then the equations in which this bit appears will no longer be satisfied, following a check on the consistency of abcd with xyz. Further, by looking at *which equations fail*, we can determine which bit is in error. Thus, the procedure is capable of correcting one error per seven transmitted bits. The cost we pay is a 3/7 reduction in the transmission rate, since only four bits carry information.

Three are there just to check consistency. In general, the Hamming code requires using n out of $2^n - 1$ bits for error checking purposes, and can correct one error out of every $2^n - 1$ bits. If there are two or more errors per group then they cannot be corrected. Obviously, the larger the value we choose for n, the more efficient is the transmission process, but the less protection one has against error.

The great advantage of the Hamming code is that the data and check bits are sent in a special order which makes decoding (error correction at receiving end, as opposed to encoding at transmission end) extremely simple. If we send abcd and xyz used in the last example in the order

$$xyazbcd \tag{8.4}$$

then we have the remarkable behaviour shown in Table 8.1. We see that the serial number of the erroneous bit contained in (8.4) is produced if we write down 1 for each check equation (8.4) that fails, O if it does not, and then form a binary number out of the three bits that we obtain in this case. The whole process is implemented easily in hardware for any value of n.

Codes with much greater error protection capabilities are available, but most of them are difficult to encode and decode, with the exception of the **Reed-Müller code**, which will guarantee correct reception as long as the number of bit errors per 2^n bits is less than $2^n/4$. This gives a much greater error-protection rate than Hamming code, especially if n is large. However, the cost is that out of 2^n bits, only n carry information. We refer the reader to texts on communication for further details.

8.5   Telemetry

8.5.1   TELEMETRY BY COMMON CARRIERS

A common carrier is a public utility with a mandate to serve the communication requirements of the population of a region. Telephone and telegraph companies are common carriers. In some nations such mandates are held by government

Table 8.1   **Decoding Hamming code**

Digit in error	Equations in error			
	3	*2*	*1*	
x	–	–	E	x appears in equation 1 only
y	–	E	–	y appears in equation 2 only
a	–	E	E	a appears in equations 1 and 2
z	E	–	–	z appears in equation 3 only
b	E	–	E	b appears in equations 3 and 1
c	E	E	–	c appears in equations 3 and 2
d	E	E	E	d appears in all the equations

departments or statutory corporations (The PTTs in Europe; Telecom in Australia), and in others by private companies with government-granted monopolies (as in the USA). Their mandate is to accept signals of a particular form, such as speech, written sheets of telegraph forms, or digital information, at a location chosen by the customer, and to reproduce the information in a suitable form at another location. Speech is input into the caller's telephone mouthpiece, and reproduced with reasonable accuracy from the earpiece of the person called. Digital data, on the other hand, require a different kind of terminal equipment, but the function of acceptance at one end and accurate reproduction at the other remains the same.

Acoustic couplers are the terminal apparatus employed for telemetry from mobile, low data rate measuring devices. They transmit data through ordinary telephones. The transducer output is first digitised, possibly error-protection bits are added, and the bit strings are used to modulate a high-frequency sound wave. The composite wave is emitted into the mouthpiece of a telephone and, depending on the actual receiving system, may be reproduced by the earpiece of the receiver telephone, detected using a microphone, and then demodulated to produce the original binary bits. Error correction may then be carried out by hardware, and, if necessary, the data can be converted back into analog form.

More frequently, the receiving system is controlled by a computer system. The electric signal received from the telephone lines may be directly demodulated by a special equipment to recover the digital information and is then processed by the computer. An important use of acoustic couplers in telemetry has been medical applications. Doctors making housecalls on patients with chronic conditions can carry instruments for measuring such signals as cardiographs, electro-encephalograms, etc., which are transmitted to a central hospital system for on-line recording and analysis. This enables patients with relatively mild conditions to recuperate at home, instead of being confined to hospital beds – an expensive undertaking.

With permanently located data sources it is possible to by-pass the telephone sets and connect modulators or demodulators directly to the telephone lines. This provides higher data rates since the restrictions imposed by the acoustic coupler–telephone set combination are no longer present. As mentioned earlier, the common carrier provides a number of choices of data rate, depending on the modem chosen, and the data link itself may be either a private leased line or a dial-up line. The former type of link is permanently established, whereas the latter type is connected upon request as with telephone calls. The user chooses his data rate and type of connection depending on his equipment, transmission times and tariff considerations, and the common carrier installs the appropriate modems and connects these to the user's devices. It is necessary for the user's equipment to conform to the electrical requirements of the modem and to send to the transmitter's data pulses of a specified rate, shape and voltage level in order that the carrier's system can correctly reproduce them at the receiver terminus. The actual way in which the system handles the signals is highly variable, depending on the distance, the data rates, the traffic density at the time the data are passing through the system, etc. They may be modulated and demodulated many times as they pass from one section to another, each of these sections having its own particular carrier and bandwidth.

As long as the common carrier's system is functioning properly, there is no *operational* difference between a remote data acquisition system and a non-remote one, since the system interface remains constant to the users. Integrated data systems may be implemented to include remote controllers, providing the correct two-way link (using full duplex lines) is installed. The extra time delay involved would have little effect in ordinary systems. However, the *design* process of the data system is complicated by the frequent need to save cost. Since it is usually cheaper to install one fast line than two lines of half the speed, multiplexing may be considered more feasible than multiple lines. The two alternatives amount to a compromise between system complexity and operational economy. Another factor is the prescribed range of data rates available from the common carrier and the need to design the system to conform with their requirements. These factors often make microprocessor-controlled equipment more favourable because of its greater flexibility over special-purpose hardware in the face of continually changing facilities offered by the common carriers.

8.5.2 SHORT-RANGE TELEMETRY BY RADIO

When transmitting data by means of a common carrier, the carrier is responsible for reproducing the data correctly at the receiving end, regardless of the distance involved, so that the range of the carrier system has little effect on the user's own system design. Where a radio link is established for telemetric purposes the situation is completely changed [12, 13].

Consider, for example, the problem of radio tracking a wild animal. A transmitter is attached to the animal, and this is set to emit a radio wave of some frequency, possibly modulated by a periodic bit pattern to make the transmission easier to recognise. However, it is impossible for us to control the direction of the radio wave once we have set the animal free. Thus, the transmitting aerial must be a *broadcasting* one so that a signal may be received regardless of the direction of transmission with respect to the receiver. This means that the emitted energy is dispersed quickly with distance, and it would be possible to track the animal only at relatively short range. Moreover, the receiving aerial must also be non-directional to some extent in order to follow the movements of the animal. A receiving aerial with a high focusing power, although more sensitive, may very well fail to locate the transmission at all. A similar problem is encountered with telemetry from a small transmitter located within the body of a medical patient. It is difficult to control the exact orientation of a miniature measuring—transmission device once this has been swallowed by the patient. However, in this case it will be possible to use a focusing receiver since we will be aware, at least approximately, of the location of the transmitter.

In both examples, circumstances place limitations on the frequency of the carrier. To transmit a radio wave efficiently, it is necessary for the size of the aerial to be comparable with the wavelength transmitted. A focusing aerial needs to be much greater. Since the transmitters used in these applications are relatively small, wavelengths in the centimetre range are usually used. There is, however, little problem to accommodate the bandwidth required. Since the carriers have frequencies of $10^9 - 10^{10}$ Hz and oscillators and receiver filters generally require passbands which increase with the carrier frequency, the bandwidth of the radio link will be quite large — much larger than normal data rate requirements. It is therefore feasible to multiplex several different measured signals before transmission. For example, devices planted inside a live body would usually measure simultaneously the temperature, fluid pressure, and concentrations of a number of chemicals, all requiring different transducers. The signals are digitised and applied to a multiplexing A—D converter, the output of which is used to modulate the carrier with the combined data stream. This may later be separated out by the receiver into the original data channels. Because of the wide bandwidth, the carrier need not possess a very exact sinusoidal form so that the oscillator and modulator are not difficult to construct and simple, low-cost devices may be used. In contrast, narrow-band systems demand more accurate design and construction, since any departure from ideal performance will alter the frequency content of the carrier and possibly remove it outside the system passband at the receiving end.

The power requirements of the small transmitters have to be calculated rather carefully in these applications. Too little power will make observations impossible, and too much will increase the size and weight of the batteries carried within the device. As a rule of thumb, we find that at a frequency of 10^8 Hz the power requirement is approximately $L^{2.5}$ watts of power radiated from the aerial, where L is the range in kilometres. Less power is needed in open country and more in thick forest.

A remark on *active* or two-way telemetry is in order here. Here, the remote device receives as well as transmits information to or from the data system. This may have a minor purpose, such as enabling observers to switch the transmitter on or off to save batteries. The receiver system in the remote device is always on and it controls the switch of the transmitter, which operates only when the observers wish to receive data. Since the transmitter uses much more power than the receiver, there is an overall saving. However, active telemetry can have much greater potential than this. Thus, in a research satellite, control telemetry would be used to initiate data transmission for specific requirements, e.g. upper-atmosphere X-ray or electron density measurements. Since information may have been gathered and recorded on the built-in tape recorder during an earlier section of the satellite's orbit, the telemetry control may cause transmission with an enhanced data transmission rate whilst the satellite is within range of a specific data receiving station.

8.5.3 LONG-RANGE TELEMETRY BY RADIO

Telemetry over long distances by radio is a large-scale operation, requiring powerful transmitters, sensitive and highly selective receivers, and specially constructed aerials. Consequently, it is employed only for corporate activities, such as ship and aircraft navigation, meteorological data transmission, or space communication.

When radio was first used in ships, long wavelengths were favoured. Such waves do not 'see' even big geological structures like mountains, and can curve along the surface of the earth with little attenuation, achieving very reliable data transmission. Short waves, in contrast, do not propagate well along the Earth's surface. However, they can be used for long-range broadcasting, because they are reflected by the ionosphere. Transmission reliability is highly dependent on the behaviour of the Sun which has major effects on the various ionospheric layers. Until recently, microwaves could not be used for long-range transmission at all, as transmitters and receivers need to be sited along a 'line of sight' transmission path having no obstruction in between. Long-range microwave systems required the construction of a series of transmission towers at regular points along the route to relay the signal and thus overcome the effects of the curvature of the Earth.

In recent years, the situation has been completely changed through the advent of the communication satellite [14, 15]. Because this is situated high above the Earth, it has a 'line of sight' over a large area, so that a transmitter—receiver pair which normally cannot communicate directly can do so via the satellite, which acts as a relay station accepting data from the transmitter and re-transmitting it for distant reception at the receiver.

As mentioned earlier, microwaves, because of their much higher frequencies, provide great bandwidths, so that a communication satellite channel operating in this way has extremely high data-carrying capacity. By employing multiplexing techniques, a ground station can pack the data of a very large number of users into one stream for transmission. Such satellite transmit—receive stations are now an integral part of the common carrier network.

The process of communication over such long ranges is a two-tier process. First there is the need simply to maintain contact. Only if this is completely satisfied is there any possibility of actually transmitting information. Whereas in ordinary systems the former is merely a minor aspect of the entire process (e.g. dialling up when we wish to make a telephone call), for satellite communication the task is more involved. It must be possible to identify and follow the position of a satellite and to direct the receiver aerial towards it, making suitable correction to the aerial position as required. Such a process is carried out concurrently with the actual data transmission.

Of particularly great importance in long-distance telemetry is error protection, since very little signal power reaches its eventual destination and interference and distortion *en route* are high. Complex coding and decoding algorithms implementing protection schemes of high error-correcting capability, make computers an indispensible part of such data transmission systems.

Long-distance telemetry is now present in everyday life and in many new scientific developments. Weather and Earth resources satellites send pictorial and other data constantly to the ground, to be shown on the home television set during the evening news. Spaceships and ground support form an integrated data system. Computer monitor and control equipment in the vehicle itself perform the local control functions to keep the machinery and instruments in order, while selected items of information are constantly relayed to ground, where decisions regarding trajectory and major operations such as landing are made.

The importance of telemetry in particular, and of data acquistion in general, is only too obvious. It is our hope that we have in this book given the reader the basis for further study into this growing subject.

References

1. SCHANDA, E. (ed.), *Remote Sensing for Environmental Sciences*. Springer-Verlag, Berlin, 1976.
2. HOLZ, R. K. (ed.), *The Surveillant Science: Remote Sensing of the Environment*. Houghton Mifflin Co., Boston, 1973.
3. BARRETT, E. C. and CURTIS, L. F. (eds.), *Environmental Remote Sensing: Applications and Achievements*. Edward Arnold, London, 1974.
4. ESTES, J. W. and SENGER, L. W. *Remote Sensing*. Hamilton, Santa Barbara (Ca., USA), 1974.
5. HARALICK, R. M., SHANMUGAN, K. and DINSTEIN, I. On some quickly computable features for texture analysis. *Proc. Symposium on computer image processing*. University of Missouri, 2, 12–2–1, 1972.
6. FITZGERALD, J. and EASON, T. S. *Fundamentals of Data Communications*. Wiley, New York, 1978.
7. BEAUCHAMP, K. G. and YUEN, C. K. *Digital Methods for Signal Analysis*. George Allen & Unwin, London, 1979.
8. GREGG, W. D. *Analog and Digital Communication*. Wiley, New York, 1977.
9. UK Post Office. *Handbook of Data Communications*. NCC Publications, Manchester, 1975.
10. DAVIES, D. W. and BARBER, D. L. A. *Communications Networks for Computers*. Wiley, New York, 1973.
11. BEAUCHAMP, K. G. (ed.), *Interlinking of Computer Networks*. Reidel, Dordrecht, 1979.
12. CACERES, C. A. (ed.), *Biomedical Telemetry*. Academic Press, New York, 1965.
13. MACKAY, R. S. *Bio-Medical Telemetry* (2nd edn). Wiley, New York, 1970.
14. MARTIN, J. *Communication Satellite Systems*. Prentice-Hall, Englewood Cliffs (NJ, USA), 1979.
15. SPILKER Jr., J. J. *Digital Communications by Satellite*. Prentice-Hall, Englewood Cliffs (NJ, USA), 1977.

INDEX